Electromagnetic Waves and Transmission Lines

Ray Kwok, Ph.D.

Copyright © 2017 by Ray Kwok

Sentia Publishing Company has the exclusive rights to reproduce this work, to prepare derivative works from this work, to publicly distribute this work, to publicly perform this work, and to publicly display this work.

All rights reserved. No part of this publication may be reproduced, stored in a retrieval system, or transmitted, in any form or by any means, electronic, mechanical, photocopying, recording, or otherwise, without the prior written permission of the copyright owner.

Printed in the United States of America

ISBN 978-0-9987874-7-3

Preface

This book is a collection of lectures in electromagnetism, waves, transmission lines, and microwave engineering given in the past 15 years at San Jose State University (SJSU). The Electromagnetic Wave was originally a one-year course for students in junior standing, and now reduced to a semester under the assumption that our students have already mastered basic electromagnetism up to Faraday's Law. In reality, many students have forgotten, and still struggling with vector algebra and applications of calculus. This is not unique to SJSU by any means, but is also true at other universities I have taught in the past, and was confirmed by other professors in many institutes across the nation.

The few textbooks we had adopted at SJSU in the past decades were either too simple (in which none of the transmission line theory was presented), or too advanced (where no adequate review in basic electromagnetism was given). This book is written for students who have learned elementary electromagnetism, but can use some review, and at the same time learn to deal with more difficult problems. Mastering problem solving skills is more important to engineering students than remembering formulas. To stimulate their logical thinking, challenging problems such as those require application of 1-dimensional result to solve 2 or 3-dimensional problems are included as examples and also as exercises at the end of the chapters.

The second half of the book focuses on waves, transmission lines, and impedance matching. Relations of the course materials to applications in antenna, RF circuits, and optics are highlighted whenever possible. Students often tell me this experience help them land their first job or internship. Many are motivated to further their study in RF and microwave communications as a result.

With only one semester, a few chapters have to be eliminated partially or entirely. Professors can pick and choose their preference. At SJSU, we tend to keep Chapters 2 to 7, and other chapters in rotation.

This textbook does not cover all aspects in electromagnetism. Many interesting topics related to the interactions between matter and electromagnetic waves are beyond what we can manage in one semester or one year. The goal of this course is to provide some fundamental knowledge of the subject, and spark enough interest for students to explore further in applied electromagnetism, material science, or RF and microwave engineering.

The author

Dr. Raymond Kwok, a Ph.D. in physics from UCLA, has been teaching physics and microwave engineering in colleges and universities for over 25 years in California. His interest in microwave engineering and antenna was inspired by the late Prof. Julien Schwinger, when Ray was taking the post-graduate level classical electrodynamic course from the Nobel Laureate. Among other achievements, Prof. Schwinger was known for leading the development of microwave and radar technology at MIT during the World War II. Ray joined San Jose State University as an adjunct faculty in 2002. For over 25 years, he has also been working as a RF engineer and consultant in satellite communications, wireless and mobile communications, RFID technologies, and electronic warfare. Dr. Kwok has over 50 publications in IEEE-MTT, Phys. Rev. Letters, and other journals, on topics of microwave engineering and solid state physics.

Electromagnetic Waves and Transmission Lines

Ray Kwok, Ph.D.

Table of Content

Chapter 1:	Vector Algebra and Vector Calculus	2
Chapter 2:	Electrostatics	13
Chapter 3:	Magnetism	43
Chapter 4:	Electrodynamics	54
Chapter 5:	Maxwell's Equations and Boundary Conditions	63
Chapter 6:	Plane electromagnetic waves	71
Chapter 7:	Normal incidence	87
Chapter 8:	Transmission Line Theory	95

Appendices

Appendix A:	Transformation of selected coordinates	121
Appendix B:	Differential operations in selected coordinates	123
Appendix C:	Trigonometric (Trig) substitution	125
Appendix D:	Wave Equations	127
Appendix E:	Standing waves	131
Appendix F:	Decibel (dB)	135

1 Vector Algebra and Vector Calculus

Understanding vectors is essential in learning electromagnetism. Readers should already be familiar with elementary vector operations. This chapter provides a brief review of the fundamentals and introduces alternative ways of describing vector operations. The benefit of using advanced notations is its simplicity when dealing with multiple-vectors product.

1.1 Vector and Scalar

Physical quantities that do not have directional information are called scalars. Examples are temperature, time, and energy. They are basically pure numbers, with physical meaning or units. Quantities that have both magnitude and direction are called vectors. Acceleration, force and momentum are examples of vectors.

1.2 Unit Vectors

There are many ways to describe the direction of a vector. In Cartesian coordinates, one might use, for example, $\hat{x}, \hat{y},$ and $\hat{z}$ to denote the vectors along the x, y and z axes with a magnitude of 1 unit. Sometimes, it is easier to use $\hat{e}_1, \hat{e}_2,$ and $\hat{e}_3$ instead, especially if we want to use matrix representation or write a computer program to do calculations. It is also convenient to use it in any other coordinate systems. For examples, $\hat{e}_1, \hat{e}_2, \hat{e}_3$ might represent $\hat{r}, \hat{\phi}, \hat{z}$ in the cylindrical coordinates, or $\hat{R}, \hat{\theta}, \hat{\phi}$ in the spherical coordinates.

1.3 Vector components

A vector component is merely the projection of a vector onto the coordinate unit vector. For examples: $A_x = \vec{A} \cdot \hat{x}$, or $B_\phi = \vec{B} \cdot \hat{\phi}$ ... etc.
Any vector can be written as a linear combination of the components in any particular coordinate system. For example, in a 2-dimensional space, $\vec{A} = 4\hat{x} - 3\hat{y}$ can also be written as $\vec{A} = 5\hat{r}$ in polar coordinate. Note that $\hat{r}$ is not a constant, unlike $\hat{x}$ or $\hat{y}$! It depends on the (x,y) coordinates of the point of interest. Only when it is mapped onto a particular Cartesian (or rectangular) coordinates would the angle shows up. In this particular case, $\hat{r} = \cos(36.9^o)\hat{x} - \sin(36.9^o)\hat{y}$, or in a compact notation, $\vec{A} = 5\angle -36.9^o$.

While x = r cosθ is the general formula to find the x-component of the vector r, there is a lot of confusion arises in practice. It is clear that the angle θ must be defined as the angle from the positive x-axis. However, the angle or the coordinates given in a problem might not be so obvious. A incline plane problem in classical mechanics is a good example.

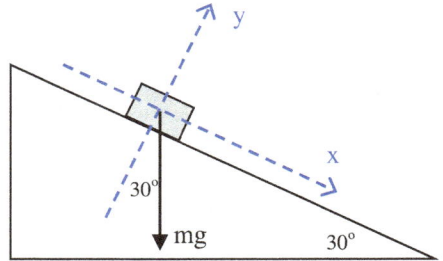

Fig.(1.1) – A incline plane problem in classical mechanics.

2

In Fig.(1.1) above, what is the x-component of the gravitational force (mg)? The answer is NOT mg(cosθ), but mg(sinθ). Some students rely on drawing a right-angle triangle to figure it out, but often make mistake on identifying the right angle. A better way is to remember the component that has the angle given in-between is the cosine component. For instance, the y-component of the gravitational force in the above example is mg(cosθ) because the angle θ is between the mg-force and the component of interest. With a little practice, people find this method easier, faster and error-free.

1.4 Addition and Subtraction

Adding or subtracting 2 vectors is easy when both vectors are represented in the same coordinate system. Explicitly, the ith component of the sum is just the sum of the ith components; $(\vec{A} \pm \vec{B})_i = A_i + B_i$.

1.5 Multiplication

When a vector multiplies with a scalar, distributive rule applies.
For example: $\vec{A} = 3\hat{x} + 4\hat{y} - 5\hat{z}$,

$$3(\vec{A}) = 3(3\hat{x} + 4\hat{y} - 5\hat{z}) = 9\hat{x} + 12\hat{y} - 15\hat{z}.$$

When a vector multiplies with another vector, there are 2 basic choices without going into the more advanced Tensor theory. They are the scalar product and the vector product.

1.5.1 Scalar Product or Dot Product

The fundamental definition of a dot product is:
$$\vec{A} \cdot \vec{B} = AB\cos\theta, \qquad \text{Eqn.(1.1)}$$

where θ is the angle between the 2 vectors. With this definition, one can easily see that:
$$\hat{x} \cdot \hat{x} = 1 = \hat{y} \cdot \hat{y} = \hat{z} \cdot \hat{z}$$
$$\hat{x} \cdot \hat{y} = 0 = \hat{x} \cdot \hat{z} = \hat{y} \cdot \hat{z}$$

Or $\hat{e}_i \cdot \hat{e}_j = \delta_{ij}$, where δ_{ij} is the Kronecker delta which has the value of 1 if i = j, and 0 otherwise. It is also the matrix elements (i^{th} row and j^{th} column) of an Identity Matrix in which all the diagonal elements are equal to 1 and the rest are zero.

Therefore, the dot product can be expressed in another way:
$$\vec{A} \cdot \vec{B} = (A_x\hat{x} + A_y\hat{y} + A_z\hat{z}) \cdot (B_x\hat{x} + B_y\hat{y} + B_z\hat{z})$$
$$\vec{A} \cdot \vec{B} = A_xB_x + A_yB_y + A_zB_z \qquad \text{Eqn.(1.2)}$$

Of course, both equations are equivalent. Equation (1.2) is more useful if the components (not the magnitude and angle) of the vectors are given. In a compact notation, we can write: $\vec{A} \cdot \vec{B} = A_i B_i$. Here, i runs from 1 to 3, representing the ith component of the vector in ANY orthogonal coordinate system. The last step, with the summation sign omitted, is a common practice for simplification and will be adopted in this chapter as well. It is implied that every time REPEATED index is used, a summation operation is there. So, another way to express the dot product is:

$$\vec{A} \cdot \vec{B} = (A_i \hat{e}_i) \cdot (B_j \hat{e}_j) = A_i B_j \hat{e}_i \cdot \hat{e}_j = A_i B_j \delta_{ij} = A_i B_i.$$

This is a very useful notation and concept when one has to deal with multiple vector products. The technique will be further discussed in later sections.

<u>The meaning of a dot product</u> $\vec{A} \cdot \vec{B}$ can be viewed as the projection of one vector $\vec{A}$, onto the other vector $\vec{B}$, and multiplies with $\vec{B}$. In other words, only the component of $\vec{A}$ vector along the vector $\vec{B}$ matters. For example, consider the work done by a force pulling a block of mass along a frictionless surface:

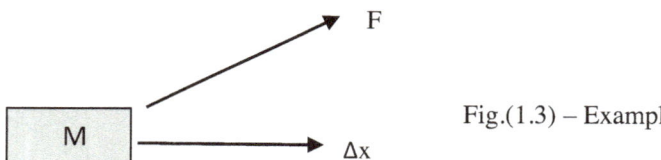

Fig.(1.3) – Example of a dot product.

Work $\equiv \vec{F} \cdot \Delta \vec{x}$, which means only the component of the force along the displacement is doing the work. The perpendicular component of the force is not doing anything at all (on a frictionless surface).

1.5.2 Vector Product or Cross Product

The cross product of $\vec{A} \times \vec{B}$ is a vector, as it is named. The magnitude is given by: $|\vec{A} \times \vec{B}| = AB \sin \theta$, and the direction is defined by the right-hand-rule.

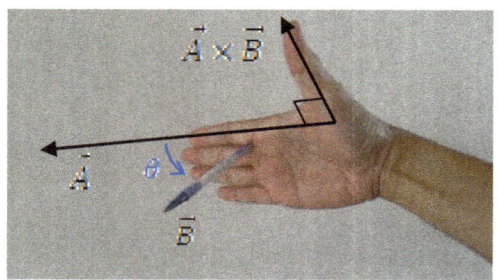

Fig.(1.4) – One way to use the right-hand-rule.

There are many ways to describe the right-hand-rule. A simple way is to follow the diagram shown. Line up the first vector $\vec{A}$ with the fingers, and have the palm facing the vector $\vec{B}$. The resultant $\vec{A} \times \vec{B}$ is given by the direction of the thumb. The resultant vector $\vec{A} \times \vec{B}$ is ALWAYS perpendicular to both vectors $\vec{A}$ and $\vec{B}$, even though the vectors $\vec{A}$ and $\vec{B}$ are not perpendicular to each other in general.

The angle θ should be defined as the angle from vector $\vec{A}$ to vector $\vec{B}$ as shown in figure. However, if the angle is greater than 90°, one can use the supplementary angle instead. i.e., in the diagram below, $|\vec{A} \times \vec{B}| = AB\sin\alpha = AB\sin\theta$. This is because of the trigonometric property that $\sin(\pi - \theta) = \sin\theta$. For dot products, this substitution CANNOT be used since $\cos(\pi - \theta) = -\cos\theta \neq \cos\theta$.

Fig.(1.5) – Supplementary angles.

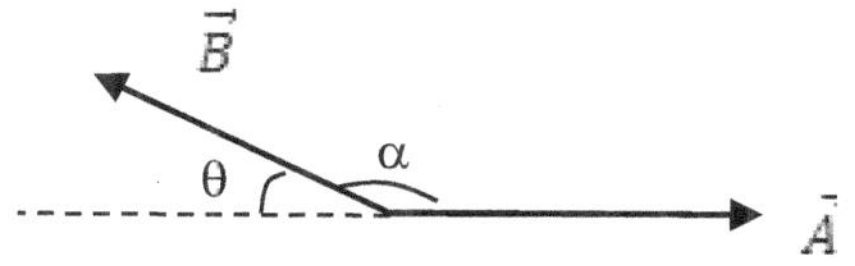

The vector notation of a cross product can be represented in determinant. For a three-dimensional vector product:

$$\vec{A} \times \vec{B} = \begin{vmatrix} \hat{e}_1 & \hat{e}_2 & \hat{e}_3 \\ A_1 & A_2 & A_3 \\ B_1 & B_2 & B_3 \end{vmatrix}$$

Eqn.(1.3)

where $\hat{e}_1, \hat{e}_2$, and $\hat{e}_3$ are the $\hat{x}, \hat{y}$, and $\hat{z}$ in Cartesian coordinates.

<u>Example 1.1</u>: Given $\vec{A} = 5\hat{x}$, and $\vec{B} = -3\hat{x} + 2\hat{y}$, (similar to the vectors shown in Fig.(1.5) above), calculate $\vec{A} \times \vec{B}$ using both methods.

(i)
$$\vec{A} \times \vec{B} = \begin{vmatrix} \hat{x} & \hat{y} & \hat{z} \\ 5 & 0 & 0 \\ -3 & 2 & 0 \end{vmatrix} = -5\begin{vmatrix} \hat{y} & \hat{z} \\ 2 & 0 \end{vmatrix} = 10\hat{z}$$

(ii)
$$|\vec{A} \times \vec{B}| = AB\sin\alpha$$
$$B = \sqrt{3^2 + 2^2} = \sqrt{13}$$
$$\alpha = \tan^{-1}\left(\frac{2}{-3}\right) = 180° - 33.7° = 146.3°$$
$$|\vec{A} \times \vec{B}| = (5)(\sqrt{13})\sin(146.3°) = 10$$

Direction is giving by the right-hand-rule, which points to $\hat{z}$, out of the page.

<u>Advanced Notation</u>

A permutation symbol ε_{ijk} is defined by:
$\varepsilon_{ijk} = 1$, if i, j, k are cyclic permutation of 1,2,3. (e.g., $\varepsilon_{123} = \varepsilon_{231} = \varepsilon_{312} = 1$)
$\varepsilon_{ijk} = -1$, if i, j, k are anti-cyclic permutation of 1,2,3. (e.g., $\varepsilon_{321} = \varepsilon_{213} = \varepsilon_{132} = -1$)
$\varepsilon_{ijk} = 0$, if any i, j, k are repeated index of 1,2,3. (e.g., $\varepsilon_{311} = \varepsilon_{223} = \varepsilon_{131} = 0$)

One can also deduce from this definition that any 2 indices can be interchanged by compensated with a negative sign. i.e., $\varepsilon_{ijk} = -\varepsilon_{ikj}$, $\varepsilon_{ijk} = -\varepsilon_{jik}$ ... etc. This property is just another way to say that interchanging any 2 rows or 2 columns in a determinant is allowed as long as it is compensated by multiplying a negative sign to the determinant.

With this, one can simply write the ith component of $\left|\vec{A} \times \vec{B}\right|_i = \varepsilon_{ijk} A_j B_k$. Remember, the summation over the repeated indices j and k are implied. For example, the x-component of $\vec{A} \times \vec{B}$ can be written explicitly as:

$\left|\vec{A} \times \vec{B}\right|_1 = \varepsilon_{123} A_2 B_3 + \varepsilon_{132} A_3 B_2$, the only non-zero terms for i = 1.

With $\varepsilon_{123} = 1$ and $\varepsilon_{132} = -1$, $\left|\vec{A} \times \vec{B}\right|_1 = A_2 B_3 - A_3 B_2$ or $\left|\vec{A} \times \vec{B}\right|_x = A_y B_z - A_z B_y$.

This can be verified easily by expanding the determinant. Again, this notation is useful when dealing with multiple vector products in later sections.

<u>The meaning of a cross product</u> $\vec{A} \times \vec{B}$ is quite opposite to $\vec{A} \cdot \vec{B}$. Specifically, it is the perpendicular component that counts, not the parallel component. It could be interpreted as the perpendicular component of A times B, or the perpendicular component of B times A. For example, torque (about the pivot point) is defined by $\vec{\tau} = \vec{r} \times \vec{F}$, can be viewed as the perpendicular component of the force times the distance, or the perpendicular distance (called moment arm) times the force.

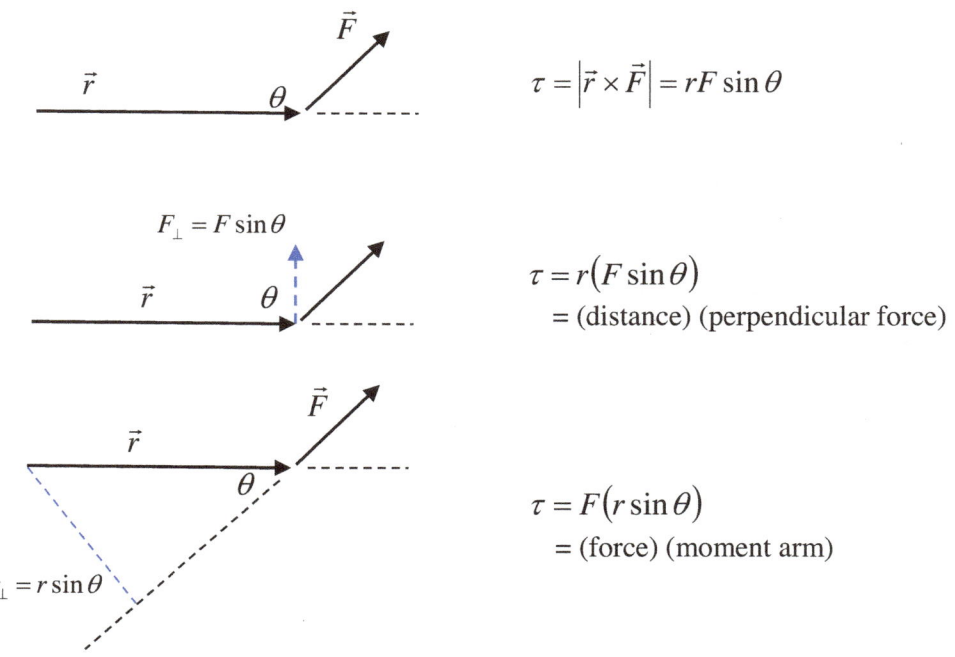

$\tau = \left|\vec{r} \times \vec{F}\right| = rF \sin\theta$

$\tau = r(F \sin\theta)$
= (distance) (perpendicular force)

$\tau = F(r \sin\theta)$
= (force) (moment arm)

1.5.3 Triple Product

There are 2 types of triple product, $\vec{A}\cdot(\vec{B}\times\vec{C})$ or $\vec{A}\times(\vec{B}\times\vec{C})$.

The first one can be evaluated easily by its representation in determinant.

$$\vec{A}\cdot(\vec{B}\times\vec{C}) = \vec{A}\cdot\vec{B}\times\vec{C} = \begin{vmatrix} A_1 & A_2 & A_3 \\ B_1 & B_2 & B_3 \\ C_1 & C_2 & C_3 \end{vmatrix}. \qquad \text{Eqn.(1.4)}$$

The parenthesis is not needed because the cross product has to be evaluated before the dot product. Using the properties of determinant, one can conclude that:

$\vec{A}\cdot\vec{B}\times\vec{C} = \vec{B}\cdot\vec{C}\times\vec{A} = \vec{C}\cdot\vec{A}\times\vec{B} = \vec{A}\cdot\vec{B}\times\vec{C} = -\vec{A}\cdot\vec{C}\times\vec{B} = -\vec{B}\cdot\vec{A}\times\vec{C}$, which is the same as the properties of the permutation symbol ε_{jk} described earlier.

<u>Example 1.2</u>: Prove that $\vec{A}\cdot\vec{B}\times\vec{C} = -\vec{B}\cdot\vec{A}\times\vec{C}$ using the ijk notation.

$$\vec{A}\cdot\vec{B}\times\vec{C} = A_i(\vec{B}\times\vec{C})_i = A_i(\varepsilon_{ijk}B_jC_k) = \varepsilon_{ijk}A_iB_jC_k = B_j(\varepsilon_{ijk}A_iC_k) = B_j(-\varepsilon_{jik}A_iC_k)$$
$$= B_j(-\vec{A}\times\vec{C})_j = -\vec{B}\cdot\vec{A}\times\vec{C}$$

To evaluate the second type of triplet product $\vec{A}\times(\vec{B}\times\vec{C})$ is very tedious using determinants over determinants. However, it becomes easier by using the advanced ijk-notation. A very useful identity to keep in mind is the product of 2 permutation symbols can be expressed in terms of products of delta functions:

$$\varepsilon_{ijk}\varepsilon_{imn} = \delta_{jm}\delta_{kn} - \delta_{jn}\delta_{km}. \qquad \text{Eqn.(1.5)}$$

Again, the summation over the repeated index "i" is implied. One can verify this relation by explicitly putting in the integers j, m, k, n, and evaluate the Kronecker delta function on the right.

Apply Eqn.(1.5) to evaluate the triple cross product $\vec{A}\times(\vec{B}\times\vec{C})$:

$$[\vec{A}\times(\vec{B}\times\vec{C})]_i = \varepsilon_{ijk}A_j(\vec{B}\times\vec{C})_k = \varepsilon_{ijk}A_j(\varepsilon_{kmn}B_mC_n) = \varepsilon_{kij}\varepsilon_{kmn}A_jB_mC_n$$
$$= (\delta_{im}\delta_{jn} - \delta_{in}\delta_{jm})A_jB_mC_n = A_jB_iC_j - A_mB_mC_i = (\vec{A}\cdot\vec{C})B_i - (\vec{A}\cdot\vec{B})C_i$$
$$\vec{A}\times(\vec{B}\times\vec{C}) = (\vec{A}\cdot\vec{C})\vec{B} - (\vec{A}\cdot\vec{B})\vec{C}.$$

Note: $\vec{A}\times(\vec{B}\times\vec{C}) \neq (\vec{A}\times\vec{B})\times\vec{C}$

$$[(\vec{A}\times\vec{B})\times\vec{C}]_i = \varepsilon_{ijk}(\vec{A}\times\vec{B})_jC_k = \varepsilon_{ijk}(\varepsilon_{jmn}A_mB_n)C_k = \varepsilon_{jki}\varepsilon_{jmn}A_mB_nC_k$$
$$= (\delta_{km}\delta_{in} - \delta_{kn}\delta_{im})A_mB_nC_k = A_kB_iC_k - A_iB_nC_n = (\vec{A}\cdot\vec{C})B_i - (\vec{B}\cdot\vec{C})A_i$$
$$(\vec{A}\times\vec{B})\times\vec{C} = (\vec{A}\cdot\vec{C})\vec{B} - (\vec{B}\cdot\vec{C})\vec{A} \neq \vec{A}\times(\vec{B}\times\vec{C}).$$

Example 1.3: $\vec{A} = 3\hat{x} - \hat{y}, \vec{B} = 2\hat{y} + 2\hat{z}, \vec{C} = -\hat{x} + 3\hat{y} - 5\hat{z}$,
Find $\vec{A} \times (\vec{B} \times \vec{C})$ and $(\vec{A} \times \vec{B}) \times \vec{C}$.

$\vec{A} \cdot \vec{C} = (3,-1,0) \cdot (-1,3,-5) = -3 - 3 + 0 = -6$
$\vec{A} \cdot \vec{B} = (3,-1,0) \cdot (0,2,2) = 0 - 2 + 0 = -2$
$\vec{A} \times (\vec{B} \times \vec{C}) = (\vec{A} \cdot \vec{C})\vec{B} - (\vec{A} \cdot \vec{B})\vec{C} = -6\vec{B} + 2\vec{C} = -6(0,2,2) + 2(-1,3,-5) = (-2,-6,-22)$
$\vec{A} \times (\vec{B} \times \vec{C}) = -2\hat{x} - 6\hat{y} - 22\hat{z}$

$\vec{B} \cdot \vec{C} = (0,2,2) \cdot (-1,3,-5) = 0 + 6 - 10 = -4$
$(\vec{A} \times \vec{B}) \times \vec{C} = (\vec{A} \cdot \vec{C})\vec{B} - (\vec{B} \cdot \vec{C})\vec{A} = -6\vec{B} + 4\vec{A} = -6(0,2,2) + 4(3,-1,0) = (12,-16,-12)$
$(\vec{A} \times \vec{B}) \times \vec{C} = 12\hat{x} - 16\hat{y} - 12\hat{z}$

One can always verify the answer using the traditional determinant method.
For example:

$$\vec{A} \times \vec{B} = \begin{vmatrix} \hat{x} & \hat{y} & \hat{z} \\ 3 & -1 & 0 \\ 0 & 2 & 2 \end{vmatrix} = \hat{x}(-2) - \hat{y}(6) + \hat{z}(6)$$

$$(\vec{A} \times \vec{B}) \times \vec{C} = \begin{vmatrix} \hat{x} & \hat{y} & \hat{z} \\ -2 & -6 & 6 \\ -1 & 3 & -5 \end{vmatrix} = \hat{x}(30 - 18) - \hat{y}(10 + 6) + \hat{z}(-6 - 6) = 12\hat{x} - 16\hat{y} - 12\hat{z}$$

Note: $\vec{A} \times (\vec{B} \times \vec{C}) \neq (\vec{A} \times \vec{B}) \times \vec{C}$. The order of cross product is not associative.

1.5.4 Multiple-Vectors Product

With the compact notation introduced in earlier sections, manipulation of multiple-vectors product becomes manageable.

Example 1.4: Show that $(\vec{A} \times \vec{B}) \cdot (\vec{C} \times \vec{D}) = (\vec{A} \cdot \vec{C})(\vec{B} \cdot \vec{D}) - (\vec{A} \cdot \vec{D})(\vec{B} \cdot \vec{C})$

$(\vec{A} \times \vec{B}) \cdot (\vec{C} \times \vec{D}) = (\vec{A} \times \vec{B})_i (\vec{C} \times \vec{D})_i = (\varepsilon_{ijk} A_j B_k)(\varepsilon_{imn} C_m D_n) = \varepsilon_{ijk} \varepsilon_{imn} A_j B_k C_m D_n$
$= (\delta_{jm} \delta_{kn} - \delta_{jn} \delta_{km}) A_j B_k C_m D_n = A_j B_k C_j D_k - A_n B_m C_m D_n = A_j C_j B_k D_k - A_n D_n B_m C_m$
$(\vec{A} \times \vec{B}) \cdot (\vec{C} \times \vec{D}) = (\vec{A} \cdot \vec{C})(\vec{B} \cdot \vec{D}) - (\vec{A} \cdot \vec{D})(\vec{B} \cdot \vec{C})$

1.6 Vector Calculus

The differentiation operator (del-operator) is defined by: $\nabla \equiv \hat{e}_i \partial_i = \hat{x}\dfrac{\partial}{\partial x} + \hat{y}\dfrac{\partial}{\partial y} + \hat{z}\dfrac{\partial}{\partial z}$

in Cartesian Coordinates. It is a vector operator with no physical meaning unless it operates on a scalar or vector function. The common operations are:

$$\nabla f = \hat{e}_i \partial_i f = \hat{x}\frac{\partial f}{\partial x} + \hat{y}\frac{\partial f}{\partial y} + \hat{z}\frac{\partial f}{\partial z} \quad \text{(Gradient)}$$

$$\nabla \cdot \vec{A} = \partial_i A_i = \frac{\partial A_x}{\partial x} + \frac{\partial A_y}{\partial y} + \frac{\partial A_z}{\partial z} \quad \text{(Divergence)}$$

$$\nabla \times \vec{A} = \hat{e}_i \varepsilon_{ijk} \partial_j A_k = \begin{vmatrix} \hat{x} & \hat{y} & \hat{z} \\ \dfrac{\partial}{\partial x} & \dfrac{\partial}{\partial y} & \dfrac{\partial}{\partial z} \\ A_x & A_y & A_z \end{vmatrix} \quad \text{(Curl)}$$

$$\nabla^2 = \nabla \cdot \nabla = \frac{\partial^2}{\partial x^2} + \frac{\partial^2}{\partial y^2} + \frac{\partial^2}{\partial z^2} \quad \text{(Laplacian)} \qquad \text{Eqn.(1.6)}$$

A del-operator applied to a scalar function is called the Gradient of that function. It usually represents the 3-dimensional "slope" of that function in a 3-dimensional space. The scalar product of the del-operator with a vector field is called the Divergence of the function. It can be thought of the outward flow or "radiation" of the vector field through some geometrical space. The Curl of a function is the vector product of the del-operator and a vector field. It takes on the meaning of the spatial "circulation" of a vector field. A Laplacian is loosely interrupted as the "divergence of a gradient" of a scalar or a vector function. It is mathematically an operator. In other words, the Laplacian of a scalar function is a scalar. If the Laplacian operates on a vector function, the final product is a vector. It is a common operator used in differential equations to describe many physical phenomena in physics and engineering. Examples in electromagnetic waves will be discussed in later chapters.

Mathematical treatment of a del-operator is similar to the vector analysis outlined above, together with all the operations associate with differentiation operator. In particular, product rule and quotient rule applied.

Example 1.5: Evaluate $\nabla \cdot (\vec{A} \times \vec{B})$.

$$\nabla \cdot (\vec{A} \times \vec{B}) = \partial_i (\vec{A} \times \vec{B})_i = \partial_i (\varepsilon_{ijk} A_j B_k) = \varepsilon_{ijk} (A_j \partial_i B_k + B_k \partial_i A_j)$$
$$= A_j \varepsilon_{ijk} \partial_i B_k + B_k \varepsilon_{ijk} \partial_i A_j = -A_j \varepsilon_{jik} \partial_i B_k + B_k \varepsilon_{kij} \partial_i A_j$$
$$= -A_j (\nabla \times \vec{B})_j + B_k (\nabla \times \vec{A})_k = \vec{B} \cdot (\nabla \times \vec{A}) - \vec{A} \cdot (\nabla \times \vec{B})$$

Example 1.6: Prove that $\nabla \times (\vec{A} \times \vec{B}) = (\vec{B} \cdot \nabla)\vec{A} + \vec{A}(\nabla \cdot \vec{B}) - \vec{B}(\nabla \cdot \vec{A}) - (\vec{A} \cdot \nabla)\vec{B}$.

$$[\nabla \times (\vec{A} \times \vec{B})]_i = \varepsilon_{ijk} \partial_j (\vec{A} \times \vec{B})_k = \varepsilon_{ijk} \partial_j (\varepsilon_{kmn} A_m B_n) = \varepsilon_{kij} \varepsilon_{kmn} \partial_j (A_m B_n)$$
$$= (\delta_{im}\delta_{jn} - \delta_{in}\delta_{jm})(A_m \partial_j B_n + B_n \partial_j A_m) = (A_i \partial_j B_j + B_j \partial_j A_i) - (A_m \partial_m B_i + B_i \partial_m A_m)$$
$$[\nabla \times (\vec{A} \times \vec{B})]_i = A_i (\nabla \cdot \vec{B}) + (\vec{B} \cdot \nabla) A_i - (\vec{A} \cdot \nabla) B_i - B_i (\nabla \cdot \vec{A})$$

$$\therefore \nabla \times (\vec{A} \times \vec{B}) = \vec{A}(\nabla \cdot \vec{B}) + (\vec{B} \cdot \nabla)\vec{A} - (\vec{A} \cdot \nabla)\vec{B} - \vec{B}(\nabla \cdot \vec{A})$$

Example 1.7: Prove that $\nabla \times (\nabla \times \vec{A}) = \nabla(\nabla \cdot \vec{A}) - \nabla^2 \vec{A}$.

$$(\nabla \times (\nabla \times \vec{A}))_i = \varepsilon_{ijk} \partial_j (\nabla \times \vec{A})_k = \varepsilon_{ijk} \partial_j \varepsilon_{kmn} \partial_m A_n = \varepsilon_{kij} \varepsilon_{kmn} \partial_j \partial_m A_n$$
$$= (\delta_{im}\delta_{jn} - \delta_{in}\delta_{jm}) \partial_j \partial_m A_n = \partial_j \partial_i A_j - \partial_m \partial_m A_i = \partial_i (\nabla \cdot \vec{A}) - \nabla^2 A_i$$

$$\therefore \nabla \times (\nabla \times \vec{A}) = \nabla(\nabla \cdot \vec{A}) - \nabla^2 \vec{A}$$

Many vector identities are used throughout the book. They will not be derived in details, but could be easily done using the techniques described in this chapter.

1.6.4 Divergence of a curl

Divergence of a curl (of any vector function) is always zero, $\nabla \cdot (\nabla \times \vec{A}) = 0$.

To prove that, one can start with Eqn.(1.4) for a triple product:

$$\nabla \cdot (\nabla \times \vec{A}) = \begin{vmatrix} \frac{\partial}{\partial x} & \frac{\partial}{\partial y} & \frac{\partial}{\partial z} \\ \frac{\partial}{\partial x} & \frac{\partial}{\partial y} & \frac{\partial}{\partial z} \\ A_x & A_y & A_z \end{vmatrix} = \frac{\partial}{\partial x}\left(\frac{\partial A_z}{\partial y} - \frac{\partial A_y}{\partial z}\right) - \frac{\partial}{\partial y}\left(\frac{\partial A_z}{\partial x} - \frac{\partial A_x}{\partial z}\right) + \frac{\partial}{\partial z}\left(\frac{\partial A_y}{\partial x} - \frac{\partial A_x}{\partial y}\right) = 0$$

The last step was achieved by expanding the determinant explicitly (as shown), or using the property that 2 rows or columns with identical entries would give a zero determinant.

1.6.5 Curl of a gradient

Curl of a gradient (of any scalar function) is always zero, $\nabla \times (\nabla f) = 0$.
One way to prove this is to expand the curl as a determinant:

$$\nabla \times (\nabla f) = \begin{vmatrix} \hat{x} & \hat{y} & \hat{z} \\ \dfrac{\partial}{\partial x} & \dfrac{\partial}{\partial y} & \dfrac{\partial}{\partial z} \\ \dfrac{\partial f}{\partial x} & \dfrac{\partial f}{\partial y} & \dfrac{\partial f}{\partial z} \end{vmatrix} = \hat{x}\left(\dfrac{\partial^2 f}{\partial y \partial z} - \dfrac{\partial^2 f}{\partial z \partial y}\right) - \hat{y}\left(\dfrac{\partial^2 f}{\partial x \partial z} - \dfrac{\partial^2 f}{\partial z \partial x}\right) + \hat{z}\left(\dfrac{\partial^2 f}{\partial x \partial y} - \dfrac{\partial^2 f}{\partial y \partial x}\right) = 0$$

1.7 Orthogonal Curvilinear Coordinates

Coordinate systems are by no means unique, and there are unlimited numbers of choices. They were invented to simplify certain types of problems. For example, it would be easier to use spherical coordinates (R, θ, φ) to describe the attraction between the Sun and the Earth, compare to using the more familiar Cartesian coordinates (x, y, z). Of course, the knowledge of transformation from one coordinate system to another is essential to obtain meaningful results. Students at this level should have already mastered this skill for the common coordinate systems. The transformation between the Cartesian coordinates, the cylindrical coordinates, and the spherical coordinates are tabulated in Appendix A.

The discussion in this chapter focused mainly on orthogonal coordinates. In other words, the unit vectors are perpendicular to each other. For example: $\hat{x}_i \cdot \hat{x}_j = \delta_{ij}$, $\hat{x} \cdot \hat{z} = 0$, or $\hat{r} \cdot \hat{\phi} = 0$. Curvilinear means the surfaces defined by the coordinates do not have to be flat, they could be curves. For example, for a constant z value, the xy-surface is a horizontal plane in Cartesian coordinates. In cylindrical coordinates, however, a constant r value gives a φz-surface that is cylindrical.

Although this book will only use the 3 special Orthogonal Curvilinear Coordinates, namely, the Cartesian, cylindrical, and the spherical coordinates, and their transformations could be easily derived from simple geometry, special care is required when they are applied to calculus. For example, an infinitesimal distance along the x-direction is dx, but that along the φ-direction in cylindrical coordinates is rdφ. Likewise, derivative along z-direction is d/dz, but the derivative along the θ-direction in spherical coordinates is not d/dθ. Explicit expressions of the differential operators in selected coordinates are summarized in Appendix B. The derivation of the transformation matrices that lead to the expressions can be found in most advanced calculus books and will not be repeated here.

Exercise:

1.1 $\vec{A} = \hat{x} - 4\hat{z}$, $\vec{B} = 2\hat{x} + \hat{y} + \hat{z}$. Find (a) $\vec{A} + \vec{B}$, (b) $\vec{B} - 2\vec{A}$, (c) $\vec{A} \cdot \vec{B}$, (d) $\vec{A} \times \vec{B}$, (e) $\vec{A} \times \vec{A}$, (f) $\vec{B} \cdot \vec{B}$, (g) the angle between $\vec{A}$ & $\vec{B}$, (h) a vector that is $\perp$ to both $\vec{A}$ & $\vec{B}$.

1.2 Prove that the volume of a right-circular cone is $\frac{1}{3}\pi r^2 h$.

1.3 A cylinder with radius R, height h, and a non-uniform mass density ρ(y) = 4 + 3y. All in SI units. If R = 1 m and h = 5 m, what is the total mass of the cylinder?

1.4 $\vec{A} = 2\hat{x} - \hat{y} + \hat{z}$, $\vec{B} = \hat{x} + 2\hat{y} - \hat{z}$, $\vec{C} = \hat{x} + \hat{y} - 2\hat{z}$. Find a unit vector in the plane of $\vec{B}$ and $\vec{C}$, and perpendicular to $\vec{A}$.

1.6 $\vec{A} = 2\hat{x} - \hat{y} + \hat{z}$, $\vec{B} = \hat{x} + 2\hat{y} - \hat{z}$, $\vec{C} = \hat{x} + \hat{y} - 2\hat{z}$. Evaluate $\vec{A} \cdot (\vec{B} \times \vec{C})$ & $\vec{A} \times (\vec{B} \times \vec{C})$.

1.7 Is this identity correct? $\vec{A} \times [\vec{A} \times (\vec{A} \times \vec{B})] \cdot \vec{C} = -|\vec{A}|^2 \vec{A} \cdot \vec{B} \times \vec{C}$. Prove it.

1.8 To prove or disprove this equation: $\vec{A} \times (\vec{B} \times \vec{C}) + \vec{B} \times (\vec{C} \times \vec{A}) + \vec{C} \times (\vec{A} \times \vec{B}) = 0$

1.9 To prove or disprove this expression: $\vec{A} \times \vec{B} = [\vec{A} \cdot (\vec{B} \times \hat{x})]\hat{x} + [\vec{A} \cdot (\vec{B} \times \hat{y})]\hat{y} + [\vec{A} \cdot (\vec{B} \times \hat{z})]\hat{z}$

1.10 $\vec{A} = x\hat{y} - xy\hat{z}$, $\vec{B} = 3\hat{x} + 4x^2 y^2 \hat{y}$. Find (a) $\vec{B} - 2\vec{A}$, (b) $\vec{A} \cdot \vec{B}$, (c) $\vec{A} \times \vec{B}$, (d) $\nabla \cdot \vec{A}$, (e) $\nabla \cdot \vec{B}$, (f) $\nabla \times \vec{A}$, (g) $\nabla \times \vec{B}$.

1.11 Prove $\nabla \cdot \nabla \times \vec{A} = 0$ explicitly in spherical coordinates.

1.12 Prove that $\nabla^2(fg) = f\nabla^2 g + g\nabla^2 f + 2\nabla f \cdot \nabla g$ for any scalar function f and g.

1.13 Prove this identity $\nabla \times (\nabla \times \vec{A}) = \nabla(\nabla \cdot \vec{A}) - \nabla^2 \vec{A}$ for any vector function $\vec{A}$.

1.14 Prove that $\nabla \times (\vec{A} \times \vec{B}) = (\vec{B} \cdot \nabla)\vec{A} + \vec{A}(\nabla \cdot \vec{B}) - \vec{B}(\nabla \cdot \vec{A}) - (\vec{A} \cdot \nabla)\vec{B}$.

1.15 Prove that $\nabla \cdot (\vec{E} \times \vec{H}) = \vec{H} \cdot (\nabla \times \vec{E}) - \vec{E} \cdot (\nabla \times \vec{H})$.

2 Electrostatics

Electrostatic was observed as early as 700 B.C. By rubbing amber with fur, ancient Greek philosophers could pick up small objects such as feathers and hair. The difference between electricity and magnetism was not understood until the 16th Century, when the term electric" was defined and systematic study of electricity began.

2.1 Electrostatic charges

At the beginning, electricity was interpreted as some kind of invisible "fluid" that carries 2 types of charges. One named "positive charge" and the other "negative". At the turn of the 20th century, the modern atomic model emerged. With that, the origin of electrostatic charges became clear.

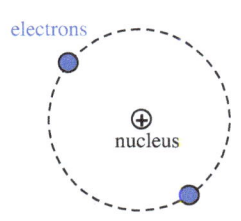

Fig.(2.1) – A Bohr atom

In the simple Bohr atomic model, electrons are orbiting around a closely packed positively-charged nucleus. There is no physical attachment between the electrons and the nucleus, so the outer electrons could easily be rubbed off. In fact, it is well-known that friction can create electrostatic charges. Most people experienced electric shock by walking on a rug and touching a metal door knob on a dry day. Workers in electronic industry wear wrist strap to ground the induced electrostatic charges, and wear anti-static smock to reduce electrostatic charges created by rubbing between fabrics.

There are many examples where electrostatic charges are revealed in daily activities. Socks stick to clothes after pull out from dryer, party balloons stuck on the wall, plastic wrap stick to ceramic plate of leftover...etc. But exactly how does it work? To understand how electric charges make things stick, one needs to take a closer look on how matters interact with charges. But what is the matter?

One way to characterize matters or materials is based on their electrical conductivity. Metals are usually good electrical and thermal conductors. Materials with poor conductivity are referred to as insulators, or "dielectric" materials. Electrons in a metal are free to move around, and only confined by the physical size of the metal. Those electrons are often referred to as "electron gas" in a box. Electrons in a dielectric can only orbit around the core atoms or molecules. These electrons react differently than the electrons in a metal when a charged object is brought nearby.

2.2 Dielectric material near a charged object

Electrons are tiny particles that move very fast. They cannot be seen or pictured as a stationery object. In fact, quantum physics describes electrons using wave functions and probability. Often, electrons are depicted as electronic "clouds", as shown in Fig.(2.2a). Since like-charges repel, opposite charges attract, a charged object nearby will polarize the neutron atom somewhat. This is illustrated in Fig.(2.2b). Even though the atom remains neutral, it is more positive on one end, and more negative on the other.

Fig.(2.2a) – A neutral atomFig.(2.2b) – A polarized atom

The polarized atom would in turn polarize other atoms around it. To what extent does this polarization line up depends on the underlying crystal structure, the molecular geometry and content. This is simplified and characterized by a material property called "dielectric constant". More discussion on dielectric properties will be presented in later chapters.

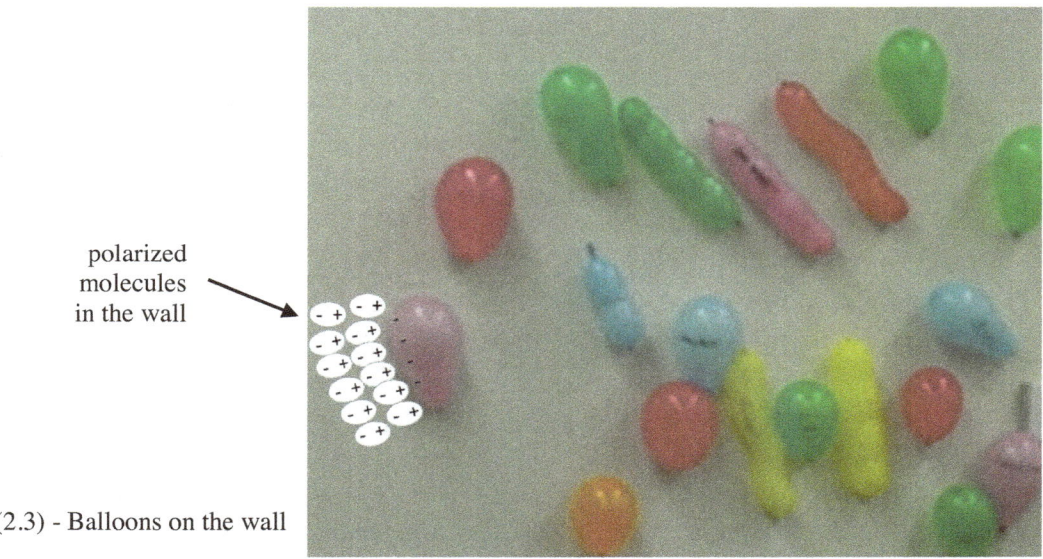

Fig.(2.3) - Balloons on the wall

When a balloon is rubbed on hair or clothing, some electrons will be rubbed off the hair and attached to the balloon. These electrons are bonded on the balloon's surface and are not free to move around. Obviously, a rubber balloon is not a conductor. When the charged balloon is brought close to a wall, which is also an insulator (dielectric material), the bonded electrons will polarize some of the molecules in the wall, as illustrated in Fig.(2.3). Since the positive end of the polarized molecules is closer to the negatively-charged balloon, the attraction force dominates. If this excess attraction force pulls the balloon close enough to the wall and creates a frictional force that can overcome the gravitational pull, the balloon will simply stick on the wall.

There are many examples of electrostatics at work. Many people use Saran Wrap, the thin clingy plastic food wrap, over ceramic plate to preserve leftover. As the plastic wrap is pulled from the roll, the friction creates unbalanced charges on the thin plastic. These bonded charges in turn polarize the plate and electrostatic force takes over. However, plastic wrap does not work on aluminum pie pans. Interactions between metals and a charged object will be examined next.

2.3 Conductor near a charged object

A primitive model of a metal is a box containing 2 types of non-interacting free charges. One type is positive, and the other is negative. They behave very much like an ideal gas and to a certain extend follow the kinetic theory of gas molecules.

Fig.(2.4) – Free charges are induced on metal by a nearby charged object

As a charged object is brought close to a conductor, free charges are induced on the surface of the conductor. Opposite attracts, and like-charges repel. Unlike the insulator, the induced charges are free to move if the path is available, such that like-charges would move as far apart as possible.

If the charged object is in contact with the conductor, the free electrons would neutralize the charged object because the electrons are free to move around. As a result, the metal is left with only positive charges, as illustrated in Fig.(2.5). And since there is no excess negative charges, there is no electrical attraction between the 2 objects. This explained why plastic food wrap does not work on metal.

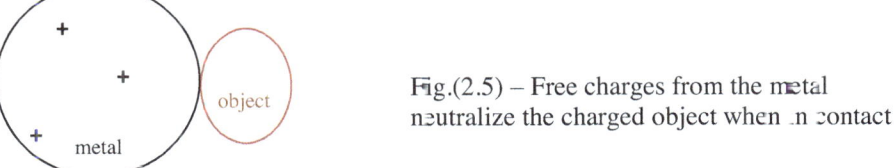

Fig.(2.5) – Free charges from the metal neutralize the charged object when in contact.

If the metal is grounded from Fig.(2.4), instead of bringing it to contact the charged object, as shown in Fig.(2.6), the positive charges will be drained. Leaving the negative charges held by the charged object. If the ground wire and the charged objects are now removed, the metal would be left with net negative charges. This process of charging the conductor without direct contact with the charged object is called "charge by induction".

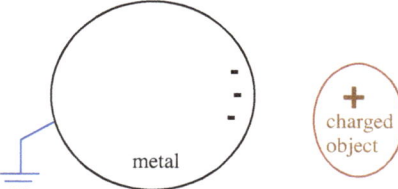

Fig.(2.6) – A ground wire drains the excess charges from the configuration in Fig.(2.4).

2.4 Coulomb's Law

The first quantitative electrical force formula was published by Coulomb in 1784:

$$F = k\frac{q_1 q_2}{r^2} \qquad \text{Eqn.(2.1)}$$

where k is the coulomb constant (9×10^9 in SI units), q_1 and q_2 are the electrical charges carried by the 2 objects (in coulombs), and r is the distance between the 2 objects. This equation is remarkably similar to Newton's universal Gravitational Force equation:

$$F = G\frac{m_1 m_2}{r^2} \qquad \text{Eqn.(2.2)}$$

where G is the universal gravitational constant (6.67×10^{-11} in SI units), m_1 and m_2 are the masses of the objects, and r is the distance between the 2 objects. Both forces obey the inverse-square law which stated the influence is inversely proportional to r^2. However, gravitational force is always attractive, while the electrical force is repulsive if the 2 charges are alike (same sign), and attractive if the 2 charges are opposite in sign (opposite attracts). And since the gravitational constant is so much smaller than the coulomb constant, the force due to gravity is expected to be much weaker than the electrical force. The following example illustrates how weak gravity is. In fact, gravity is the weakest natural forces among the 4 known forces (gravity, electromagnetic, weak and strong nuclear forces).

Example 2.1: A simple hydrogen atom consists of an electron orbiting around a proton with an average distance of 1Å. Calculate the electrical force and the gravitational force between them. [Mass of electron = 9.1×10^{-31} kg, mass of proton = 1.67×10^{-27} kg, one electronic charge = -1.6×10^{-19} C, 1Å = 10^{-10} m.]

$$F_e = k\frac{q_1 q_2}{r^2} = (9 \cdot 10^9)\frac{(1.6 \cdot 10^{-19})^2}{(10^{-10})^2} = 2.3 \cdot 10^{-9} \text{ N}$$

$$F_G = G\frac{m_1 m_2}{r^2} = (6.67 \cdot 10^{-11})\frac{(9.1 \cdot 10^{-31})(1.67 \cdot 10^{-27})}{(10^{-10})^2} = 10^{-47} \text{ N}$$

$$F_e \gg F_G$$

Fig.(2.7) – Gecko and the hair-like structure on its toes.

Gecko is a family of lizards that do not have suction cups. Instead, they have hair-like structure called setae on their toes. By rubbing them against each other, they can produce up to 290 lbs of Van de Waals (electrostatic) attractive force if all the setae are in contact with the wall.

Example 2.1 shows that gravity is basically negligible compare to the electrical force. Why then, Newton discovered gravity about 100 years before Coulomb defined electrical force? That is because macroscopic objects, such as planets and apples, are primarily neutral objects, so that there is no net electrical force between them.

2.5 Multiple point charges

Coulomb Law only applies to 2 ideal point charges, which means these charges are infinitesimally small. For any finite charge distribution, one would need calculus to break down the problems. Details and examples will be given in later sections. Just like other forces, Coulomb forces are vectors and must be added vectorially.

Example 2.2: A positive charge +2μC is located at the origin of the xy-plane. A negative -5μC is placed at 0.5m along the +x axis. What is the electrical force acting on an electron at (0, 0.2) m along the y-axis?

$$F_1 = F_{1y} = k\frac{q_1 q_2}{r^2} = -k\frac{e(2)}{(0.2)^2} = -200ke\,[\mu N]$$

$$F_{2x} = -k\frac{e(5)}{r^2}\sin\theta = -k\frac{e(5)(0.5)}{\left[(0.2)^2 + (0.5)^2\right]^{3/2}} = -16ke\,[\mu N]$$

$$F_{2y} = k\frac{e(5)}{r^2}\cos\theta = k\frac{e(5)(0.2)}{\left[(0.2)^2 + (0.5)^2\right]^{3/2}} = 6.4ke\,[\mu N]$$

$$\vec{F}_{total} = -(16\hat{x} + 194\hat{y})ke\,[\mu N] = -(2.3\hat{x} + 27.9\hat{y}) \cdot 10^{-8}[N] = 2.8 \cdot 10^{-7} \angle -95°\,[N]$$

The answer can be written in Cartesian or polar coordinates as shown. The angle in the polar form is defined from the +x axis.

Suppose instead of the electron, an alpha particle is placed at the same location (0, 0.2)m along the y-axis. What would the electrical force this particle experience? Rather than repeating the same procedure over and over again, it would be useful to introduce the concept of Electric Field.

2.6 Electric Field

Coulomb's Law describes the interaction between 2 point charges. But how do they know the existence of each other across a distance? Classical field theory provided an explanation on such encounter. In short, a positive charge radiates a field around it. When another charge encounters this field, it experiences and reacts to it as described by the Coulomb's Law. How does a charge particle radiate electric field and energy indefinitely is beyond the scope of the classical theory. It is addressed by quantum theories and will not be discussed in this book.

If the electrical force on a point charge (or often referred to as "test charge") is known, the electric field is simply defined by:

$$\vec{E} = \frac{\vec{F}}{q} \qquad \text{Eqn.(2.3)}$$

In other words, the direction of electric field is the as the direction of Coulomb force if a small test charge is placed at that location, and the magnitude is given by F/q. Knowing the direction of the Coulomb force, one can conclude that electric field originates from a positive charge and terminates into a negative charge.

<u>Example 2.3</u>: To address the follow-on question in Example 2.2, if the electron is replaced by an alpha particle (which is the nucleus of a helium atom, consists of 2 protons and 2 neutrons), how much force would it experience?

Since we know the electrical force on the electron at (0, 0.2) m already, we can calculate the electric field at that location by using Eqn.(2.3).

$$\vec{E} = \frac{\vec{F}_{total}}{q} = \frac{(16\hat{x} + 194\hat{y})ke[\mu N]}{-e} = -(16\hat{x} + 194\hat{y})k[\mu N/C]$$

$$\vec{F}_{alpha} = q\vec{E} = (+2e)[-(16\hat{x} + 194\hat{y})k] = -2[(16\hat{x} + 194\hat{y})ke]$$

$$\vec{F}_{alpha} = -(4.6\hat{x} + 55.8\hat{y}) \cdot 10^{-8}[N] = -5.60 \cdot 10^{-7} \angle 85.3°[N]$$

The concept of electric field is very useful especially when dealing with complicated charge distribution. Once the electric field is determined, electrical force can be easily evaluated. The next few sections outline steps to calculate electric field for some simple charge distributions.

2.6.1 Electric Field of point charges

For a single point charge Q, Eqns.(2.1) and (2.3) simply read:

$$\vec{E} = \frac{\vec{F}}{q} = k\frac{Q}{r^2}\hat{r} \qquad \text{Eqn.(2.4)}$$

The direction of the electric field is included in Eqn.(2.4) such that if Q is positive, the field is pointing in the positive radial direction. For multiple point charges, superposition applies, and the electric fields are added up vectorially the same way Coulomb forces are as illustrated in Example (2.2).

2.6.2 Electric Field of a continuous charge distribution

Calculus was invented to deal with this kind of problems. In particular, integration is a technique for breaking down a continuous structure into infinitesimally small elements with which simple physical rules can apply. For example, finding a volume of a continuous structure requires breaking it down into small elements with known volumes such as cubes or discs, and integrate them together. This is a very powerful tool for solving this type of problems.

Example 2.4: Find the electric field of an infinite line charge of uniform linear charge density λ [in coul/m], at a distance x away from the line.

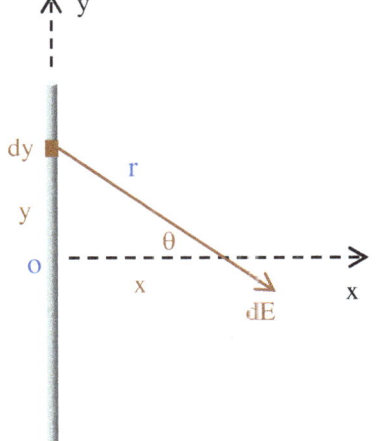

Draw a diagram to represent the problem on hand.

Label clearly all the variables and coordinates.

Draw an infinitesimal element for integration later. This typical element should not be at the center or at the ends.

Treat this infinitesimal element as a point charge and apply Eqn.(2.4), which should read: $d\vec{E} = k\frac{dq}{r^2}$

dq is the charge of the infinitesimal element = λdy

Remember dE is a vector, so it has to be separated into x and y components before adding up (or integrate). Since this is an infinite line charge, point o in the diagram can be considered as the mid-point of the line. There is infinite amount of charges above point o and infinite amount of charges below that.

Note that the net electric field points to the right of the diagram due to its symmetry. The vertical component is cancelled by the contribution of the opposite half of the line. Always look for symmetry to simplify the problem. Fail to recognize this is not a big deal, it only means there are more unnecessary integrals to do. And the calculated y-component in this case would just be zero.

$$dE = k\frac{dq}{r^2} = k\frac{\lambda dy}{r^2}$$

$$dE_x = dE\cos\theta = \left(k\frac{\lambda dy}{r^2}\right)\left(\frac{x}{r}\right) = \frac{k\lambda x dy}{r^3}$$

$$E_x = k\lambda x \int_{-\infty}^{\infty}\frac{dy}{(x^2+y^2)^{3/2}} = k\lambda x \int\frac{x\sec^2\theta d\theta}{(x\sec\theta)^3} = \frac{k\lambda}{x}\int\cos\theta d\theta = \frac{1}{x^2}\sin\theta = \frac{k\lambda}{x}\left[\frac{y}{\sqrt{x^2+y^2}}\right]_{y=-\infty}^{+\infty} = \frac{2k\lambda}{x}$$

Most of the integrals in electromagnetism depend on r, and often require trigonometric substitution. A brief review on Trig substitution is covered in Appendix C.

In most literatures, the Coulomb constant is further defined as:

$$k = \frac{1}{4\pi\varepsilon_o} \qquad \text{Eqn.(2.5)}$$

where ε_o is the free space permittivity and has a value (in SI unit) of

$$\varepsilon_o \approx \frac{10^{-9}}{36\pi} = 8.85 \cdot 10^{-12}. \qquad \text{Eqn.(2.6)}$$

Permittivity is a parameter to describe the ability a substance has to trap electric field or to store electric energy. More details on this parameter will be discussed in later sections.

With this, one can rewrite the result for the electric field of an infinite line charge in Example (2.4):

$$\vec{E}_{line} = \frac{2k\lambda}{x} = \frac{\lambda}{2\pi\varepsilon_o x}\hat{x} \qquad \text{Eqn.(2.7)}$$

Again, this formula is only valid if the line charge is infinite, and the charge distribution (density) is uniform. Otherwise, the charge density function needs to be inside the integral and the limits of the integral have to be adjusted.

2.6.3 Electric Field of an infinite charge plane

Suppose the xy-plane is uniformly charged with a surface charge density σ. What is the electric field at a distance z along the z-axis?

There are many ways to solve this problem. Some of them will be left as exercises for the readers. Two different methods will be discussed here to illustrate different problem solving strategies.

Method 1: Polar (cylindrical) coordinates

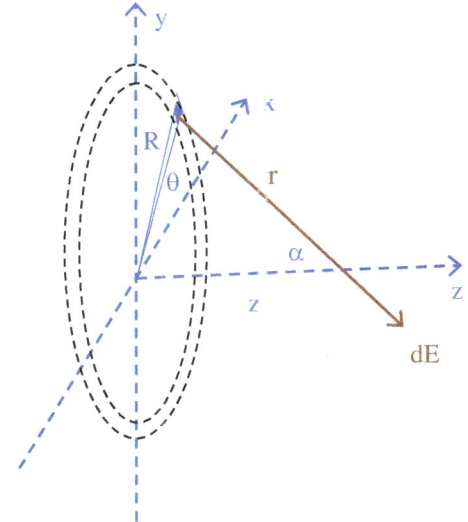

The infinitesimal element is the little section bound by R and (R + dR), θ and (θ – dθ) as drawn in figure. The "tiny" electric field (dE) from this "tiny" typical element is written as

$$dE = \frac{k(dq)}{r^2} = \frac{k(\sigma dA)}{r^2} = \frac{k\sigma(RdRd\theta)}{r^2}$$

Integrate this over θ from 0 to 2π gives the electric field of a ring. Summing up all the rings with R from 0 to infinity covers the whole infinite xy-plane. But, remember electric field is a vector. It has to be decomposed into components before integration.

Looking at the charge distribution, one can see the net electric field must be in the z-direction. The other 2 components should sum up to be zero. Therefore,

$$dE_z = dE \cos\alpha = \frac{k\sigma(RdRd\theta)}{r^2} \cdot \frac{z}{r}$$

$$E_z = k\sigma z \int_0^\infty \frac{RdR}{r^3} \int_0^{2\pi} d\theta = 2\pi k\sigma z \int_0^\infty \frac{RdR}{r^3} = 2\pi k\sigma z \int \frac{(z\tan\alpha)(z\sec^2\alpha d\alpha)}{(z\sec\alpha)^3} = 2\pi k\sigma \int_0^{\pi/2} \sin\alpha\, d\alpha = 2\pi k\sigma = \frac{\sigma}{2\varepsilon_o}$$

Again, trigonometric substitution was used in the last line to arrive to the answer. Students should repeat this calculation until the reason behind each step is fully understood.

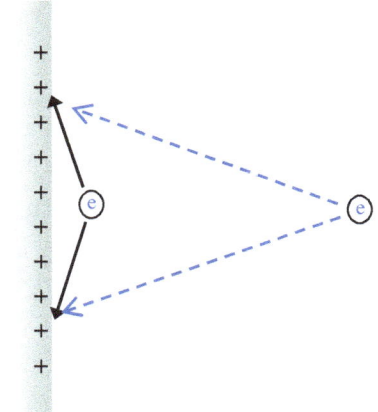

Fig.(2.7) – An uniformly charged non-conducting infinite plate

The answer to this example is also interesting. It stated that the electric field due to an infinite uniformly charged plate is equal to $\sigma/2\varepsilon_o$ which is a constant. It does not depend on the distance or the location! How could that be possible? So an electron placed at 1 mm in front of this charge-plate would experience the same electrical force as it is at 1 km away?

Yes. The answer lies on the fact that electric field is a vector. It is easy to see that the net attraction force on the electron due to this infinite charge plate is horizontally, and to the left. Therefore, only the horizontal component counts.

When an electron is very close by, the magnitude of the attraction force due to any particular charge on the plate is stronger. But just a little bit away from the center, the electrical force vectors are pointing mostly vertical, with only very little horizontal component. So the nearby charges do not contribute to the net force very much. Whereas, when the electron is far away, the force magnitude between each charge pair is weaker. But these force vectors are pointing mostly horizontally, and thereby contributing more to the overall attraction force.

The fact that these 2 effects: distance and angle compensate each other is no accident. This is expected as Coulomb Law is an inverse-square law. The mathematical proof of this is simply to repeat the calculation as presented in this example. Interested readers should also look up the concept of "solid angle".

Method 2: Make use of the result of an infinite line charge in Example (2.4)

An infinite plane is made of a set of infinite lines along a direction. So, knowing the electric field of an infinite line charge, one can find the electric field of an infinitely charged plate by integrating them using calculus.

From Eqn.(2.7), $\vec{E}_{line} = \dfrac{\lambda}{2\pi\varepsilon_o x}\hat{x}$ where λ is the linear charge density of the line. But in this example, there is no λ. Instead, it is the surface charge density σ that is given. And what is the coordinate x means?

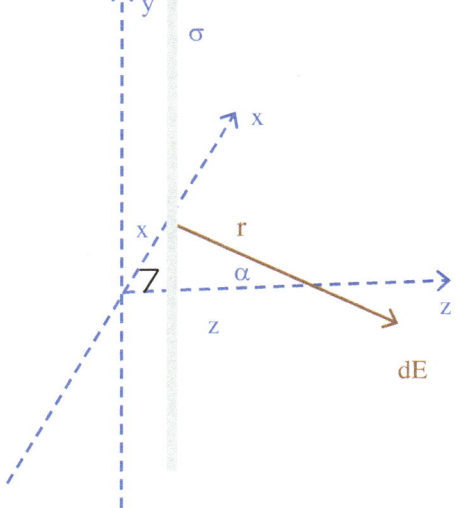

For this integral, the typical element is a thin vertical line with width dx. The "tiny" electric field due to this "tiny" element (thin) is: $dE = \dfrac{\lambda}{2\pi\varepsilon_o r}$

Note: the "x" in Eqn.(2.7) is the distance "r" in the diagram. What about λ?

In Example (2.4), the total charge on the line is λL, where $L \to \infty$. In this figure, the total charge on the typical element is $\sigma dA = \sigma L dx$. Equating them:

$\lambda L \to \sigma L dx$

$\lambda \to \sigma dx$

$dE = \dfrac{\sigma dx}{2\pi\varepsilon_o r}$

Again, because the electric field is a vector, it has to be decomposed into components before integration. As discussed before, the net field is along the z-direction.

$$dE_z = dE\cos\alpha = \left(\frac{\sigma dx}{2\pi\varepsilon_o r}\right)\left(\frac{z}{r}\right) = \frac{\sigma z}{2\pi\varepsilon_o}\frac{dx}{r^2}$$

$$E_z = \frac{\sigma z}{2\pi\varepsilon_o}\int_{-\infty}^{\infty}\frac{dx}{r^2} = \frac{\sigma z}{2\pi\varepsilon_o}\int\frac{(z\sec^2\alpha\, d\alpha)}{(z\sec\alpha)^2} = \frac{\sigma}{2\pi\varepsilon_o}\int_{-\pi/2}^{\pi/2}d\alpha = \frac{\sigma}{2\pi\varepsilon_o}(\pi) = \frac{\sigma}{2\varepsilon_o}$$

Here, the same answer is derived. This example illustrates how to integrate a partial solution of a smaller subset into a bigger problem. This is the kind of critical problem solving skills students should master and apply to other subjects and their future engineering careers.

2.6.4 Non-uniform charge distribution

The procedure to calculate electric field would be the same regardless what the charge distribution is. Very often, the charge density is given as a function of location, which would end up inside the integrals, whereas in the past, the constant charge density was put in front of the integral.

There is a more important effect that needs attention. Depends on the details of the charge density function, the charge symmetry could be altered significantly. The direction of the net electric field could be different than in the case of uniform charge distribution. The following is an example.

Example 2.5: Find the electric field of a line charge of length L along the y-axis, with linear charge density $\lambda = 2y$ [in coul/m], at a distance x away from the center of the line.

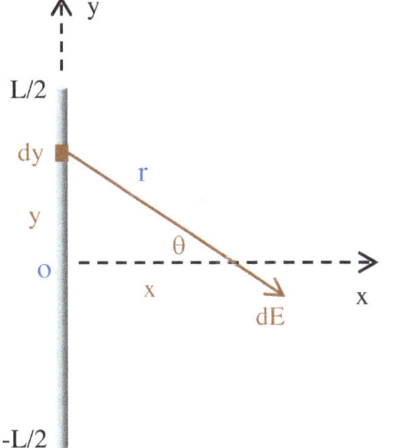

This example is very similar to Example 2.4 with a few differences.

The charge density $\lambda = 2y$ means positive charges are above the origin, and negative charges below. And they are non-uniform. Looking at this distribution, one can see the NET electric field must be pointing down along the –y direction.

As usual, we start by drawing the typical element and writing the "tiny" E-field for it:

23

$$dE = k\frac{dq}{r^2} = k\frac{\lambda dy}{r^2}$$

$$dE_y = -dE\sin\theta = -\left(k\frac{\lambda dy}{r^2}\right)\left(\frac{y}{r}\right) = -\frac{k2y^2 dy}{r^3}$$

$$E_y = -2k\int_{-\infty}^{\infty}\frac{y^2 dy}{r^3} = -2k\int\frac{(x\tan\theta)^2(x\sec^2\theta d\theta)}{(x\sec\theta)^3} = -2k\int\frac{\sin^2\theta}{\cos\theta}d\theta = -2k\left[\int\sec\theta d\theta - \int\cos\theta d\theta\right]$$

$$\int\sec\theta d\theta = \ln|\sec\theta + \tan\theta| = \ln\left|\frac{r}{x} + \frac{y}{x}\right| = \left[\ln\left|\frac{\sqrt{x^2+y^2}+y}{x}\right|\right]_{y=-L/2}^{+L/2} = \ln\left|\frac{\sqrt{x^2+L^2/4}+L/2}{\sqrt{x^2+L^2/4}-L/2}\right|$$

$$\int\cos\theta d\theta = \sin\theta = \left[\frac{y}{\sqrt{x^2+y^2}}\right]_{y=-L/2}^{+L/2} = \frac{L}{\sqrt{x^2+L^2/4}}$$

$$\vec{E} = -\hat{y}2k\left[\ln\left|\frac{\sqrt{x^2+L^2/4}+L/2}{\sqrt{x^2+L^2/4}-L/2}\right| - \frac{L}{\sqrt{x^2+L^2/4}}\right] = -\hat{y}2k\left[\ln\left|\frac{\sqrt{4x^2+L^2}+L}{\sqrt{4x^2+L^2}-L}\right| - \frac{2L}{\sqrt{4x^2+L^2}}\right]$$

2.7 Gauss's Law

In principle, Coulomb's Law together with calculus provide a way to find electric field of a charge distribution. The details calculation, however, could be quite cumbersome even for a simple 3-dimensional structure. Alternatively, one can use Gauss's Law to calculate electric field especially for simple and symmetrical charge distribution.

Gauss's Law of electricity is stated as: $\oint\vec{E}\cdot d\vec{a} = \frac{Q_{inside}}{\varepsilon_o}$ Eqn.(2.8)

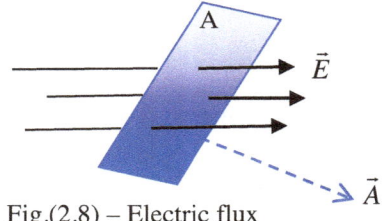

Fig.(2.8) – Electric flux through a plane.

The left side of Eqn.(2.8) is the net electric flux that goes through a volume defined by the enclosed area in the closed integral. A vector flux in mathematics is the dot product of a vector onto an area, as illustrated in Fig.(2.8). The area vector is defined as the normal (perpendicular) to surface, pointing outward away from the volume.

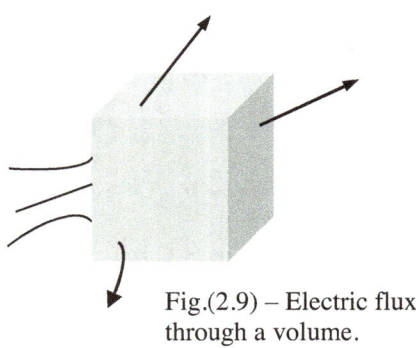

Fig.(2.9) – Electric flux through a volume.

Gauss's Law simply says that the net electric flux through a volume is proportional to the net charge stored inside the volume. In other words, if there is no net charge in the volume, the amount of electric field enters into this region has to leave with no degradation. Sometimes people refer electric field as a conservative field.

Gauss's Law is a very general theory that can be applied to enclosed surfaces of all size and shape. However, it is only useful for calculating electric field when there is a nice symmetrical charge distribution. Imagine if the area chosen for integration is such that the electric field is uniform on the surface, and $\vec{E} \parallel d\vec{a}$, then Eqn.(2.8) becomes a simple algebraic equation:

$$EA = \frac{Q_{inside}}{\varepsilon_o}.$$
Eqn.(2.9)

The area A that satisfies the 2 conditions above is called a Gaussian surface. It has to have the same symmetry as the charge distribution. There are only a few limited cases where Gaussian surfaces can be identified. A few of them will be illustrated here.

Example (2.6): A point charge.

For a single point charge Q, what is the electric field at a distance r away? Because of its total symmetry, the Gaussian surface would be just a sphere. Any point of the spherical surface has the same electric field intensity because of symmetry. The direction of E is pointing in the radial direction, so is the differential area vector.

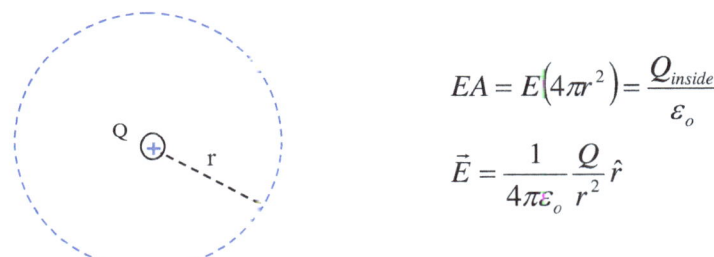

$$EA = E(4\pi r^2) = \frac{Q_{inside}}{\varepsilon_o}$$

$$\vec{E} = \frac{1}{4\pi\varepsilon_o}\frac{Q}{r^2}\hat{r}$$

The result is, of course, just the Coulomb equation for a point charge.

Example (2.7): An infinite line charge with uniform charge density λ.

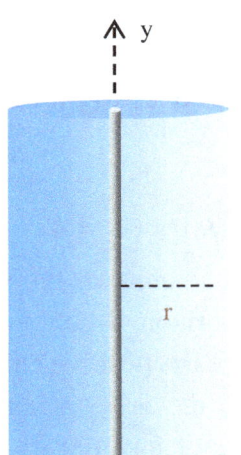

$$EA = E(2\pi rL) = \frac{Q_{inside}}{\varepsilon_o} = \frac{\lambda L}{\varepsilon_o}$$

$$\vec{E} = \frac{\lambda}{2\pi\varepsilon_o r}\hat{r}$$

which is the same as Eqn.(2.7) in Example (2.4).
Note that if the line length is finite as in Example (2.5), no simple Gaussian surface can be identified. The electric field intensity near the far end of the line charge would be quite different than around the center. Even if a surface can be identified to satisfy the 2 criteria to be a Gaussian surface, t would not be a simple geometry and the area would not be readily available. Hence, Eqn.(2.9) would not be useful.

Likewise, if the charge density is not uniform, Gaussian surface would not be easily identified or calculated.

Example (2.8): An infinite charged plate with uniform charge density σ.

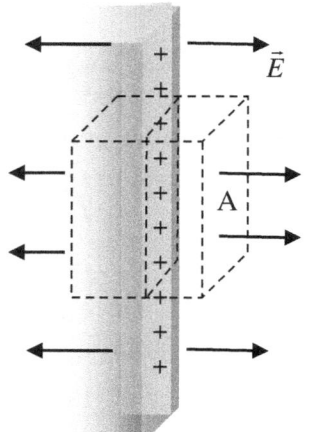

Here, the Gaussian surface is a thin box with cross-sectional area A parallel to the infinite plate. The shape of A is arbitrary as long as it is parallel to the plate. Only the front and the back surface (with area A) contribute to the flux. The other 4 faces have the area vector perpendicular to the E-field, so $\vec{E} \cdot \vec{A} = 0$.

$$\oint \vec{E} \cdot d\vec{a} = EA + EA = \frac{Q_{inside}}{\varepsilon_o} = \frac{\sigma A}{\varepsilon_o}$$

$$E = \frac{\sigma}{2\varepsilon_o}$$

Again, the same electric field is obtained as in Section 2.6.3.

2.8 Electrostatic charge on conductor

When free charges are deposited on or induced in metal, where do they go? How to find the electric field generated around them? It would be useful to have a few of the following properties in mind when working with free charges on metal.

2.8.1 No electrostatic field inside a conductor.

Without applying an external power source, there is no electric field inside a conductor. Imagine if the electric is not zero, there will be electrical force accelerating charges inside the conductor according to Eqn.(2.3). Therefore, charges will be moving and creating an electrical current when there is no external power. This just cannot be true.

2.8.2 Free charges only stay on the surface of a conductor.

No electric field inside also means no net charges inside a conductor as dictated by Gauss's Law Eqn.(2.8). Therefore, any free moving charges introduced to a conductor must stay as surface charges only.

2.8.3 Electric field on the surface of the conductor is perpendicular to the surface.

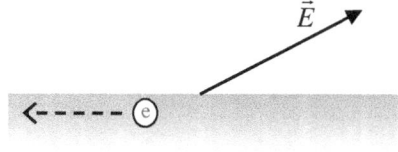

Fig.(2.10) – If electric field is NOT perpendicular to the surface of a conductor….

If the electric field is not perpendicular to the surface, as shown in figure, then the horizontal component of the electric field is going to drive the positive charges to the right, and free electrons to the left. Again, there will be a surface current without any power source. This cannot happen. So electric field on the conductor surface has to be perpendicular to the surface.

2.8.4 Electrostatic charges distribute uniformly only on perfectly symmetrical surfaces.

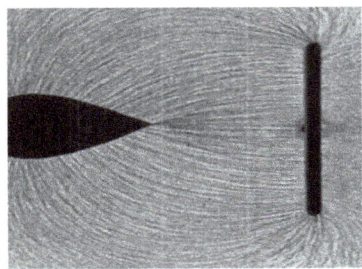

Fig.(2.11) – Electric Field from sharp charged objects.

One of the experiments in freshman physics classes was to map the electric field of various charged conductors. Electric field is more intense around sharp corners. Knowing that electric field originates from positive charges, this also means more free charges cumulate at sharp corners.

Remember a simple cube have sharp corners too. It is wrong to assume free charge distribution on the surface of a metallic box is uniform.

<u>Example (2.9)</u>: A conducting sphere with radius 'a' and induced charges Q, is shielded by a spherical conducting shell of inner radius 'b' and outer radius 'c'. The outer shell is grounded as shown in the figure. The space between the sphere and the shell is filled with air only. Find electric field everywhere in space. Plot E(r).

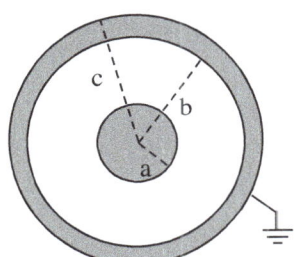

For r < a, E = 0 (inside a conductor).
All the charges stay on the surface r = a.
Surface charge density at r = a is $\sigma = \dfrac{Q}{4\pi a^2}$

For b > r > a, $EA = \dfrac{Q_{total}}{\varepsilon_o}$

$$E(4\pi r^2) = \dfrac{Q}{\varepsilon_o}$$

$$\vec{E} = \dfrac{Q}{4\pi \varepsilon_o r^2}\hat{r}$$ As if all the charges is concentrated at the origin, like a point charge.

For c > r > b, E = 0 (inside a conductor).
That means net charge inside (r < b) must be zero, according to Gauss's Law.
Total charge induced on the inner surface of the shell at r = b must be = -Q.
And the surface charge density at r = b is $\sigma = \dfrac{-Q}{4\pi b^2}$

For r > c, E = 0 (grounded and shielded.)

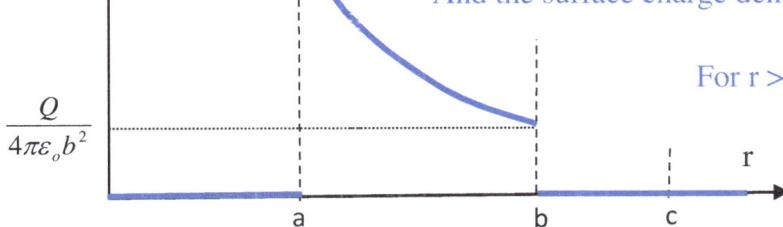

2.9 Electrostatic charge on dielectric

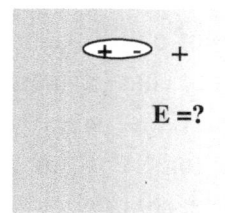

Fig.(2.12) – A free charge in dielectric induced polarization.

When charges are introduced in dielectric, some surrounding molecules will be polarized. Electric field in the dielectric would be different due to the induced electric dipole, although the net charge is still the same. Calculating electric field in dielectric becomes a complicated task. The concepts of polarization charge and displacement vector were introduced to simplify the problem.

2.9.1 Polarization charge

The right-hand-side of Eqn.(2.8) is proportional to the total charge enclosed in the volume bounded by the surface integral on the left-hand-side. Total charge here includes not only the free charges, but also the induced polarized molecules. In Fig.(2.12), the electric field in the dielectric is certainly different if the polarized molecule is not there. The difference is considered as the contribution of polarization charges or bound charges. Therefore, the total charge is the sum of the free charge and the polarization charges.

$Q_{total} = Q_{free} + Q_{bound}$. Eqn.(2.10)

More discussion on Gauss's Law and polarization charges will be included in a later chapter of Maxwell's Equations.

2.9.2 Displacement Vector and dielectric constant

To separate the contribution from free charges and polarized charges, a displacement vector is introduced. For the free charges contribution, the Gauss's Law reads as:

$$\oint \vec{D} \cdot d\vec{a} = Q_{free}$$ Eqn.(2.11)

Here, Q_{free} is the free charges enclosed by the enclosed surface. The displacement vector D can be approximated as:

$$\vec{D} = \varepsilon \vec{E} = \varepsilon_o \varepsilon_r \vec{E}$$ Eqn.(2.12)

where ε_r is the relative dielectric constant of the medium. It is a parameter to indicate how easily the dielectric material can be polarized by nearby charges. It also reveals the amount of electric field or electric energy can be stored in the material.

Eqn.(2.12) suggests that the displacement vector D is along the same direction as the electric field vector E. Apparently, this is not exactly correct, as illustrated in Fig.(2.12). However, at a distance away from the dipole, Eqn.(2.12) is a good approximation. Without getting too much into material science, crystal structure and tensor algebra, Eqn.(2.12) is valid in general, and dielectric constant is an over-simplified but effective way to describe the electric property of the material. When dealing with dielectric material, Eqn.(2.11) should be a starting point in problem solving.

Example (2.10): A very long coaxial cable is made of a dielectric with a radius of "a" and a relative dielectric constant ε_r. The outer shield is a perfect conductor with inner radius "b" and outer radius "c", as shown in figure. Space between "a" and "b" is just filled with air. A non-uniform charge density $\rho = \alpha r$ is distributed in dielectric where r < a. Find the electric field everywhere. Sketch it.

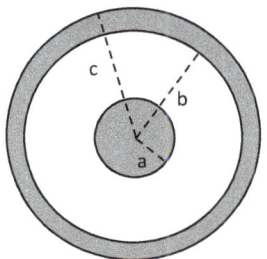

For r < a,
choose a long cylinder as the Gaussian surface in the dielectric

$$\oint \vec{D} \cdot d\vec{a} = Q_{free}$$

$$DA = \int_0^r \rho dV = \int_0^r (\alpha r)(2\pi r L dr) = 2\pi \alpha L \int_0^r r^2 dr$$

$$\varepsilon_o \varepsilon_r E(2\pi r L) = \frac{2\pi \alpha L}{3} r^3$$

$$\vec{E} = \frac{\alpha}{3\varepsilon_o \varepsilon_r} r^2 \hat{r}$$

For b > r > a, in air

$$\oint \vec{D} \cdot d\vec{a} = Q_{free}$$

$$DA = Q_{inside} = \int_0^a \rho dV = \frac{2\pi \alpha L}{3} a^3$$

$$\varepsilon_o E(2\pi r L) = \frac{2\pi \alpha L}{3} a^3$$

$$\vec{E} = \frac{\alpha a^3}{3\varepsilon_o r} \hat{r}$$

For c > r > b, in conductor, E = 0.

For r > c, outside the cable,

$$\oint \vec{D} \cdot d\vec{a} = Q_{free}$$

$$DA = Q_{inside} = \int_0^a \rho dV = \frac{2\pi \alpha L}{3} a^3$$

$$\varepsilon_o E(2\pi r L) = \frac{2\pi \alpha L}{3} a^3$$

$$\vec{E} = \frac{\alpha a^3}{3\varepsilon_o r} \hat{r}$$

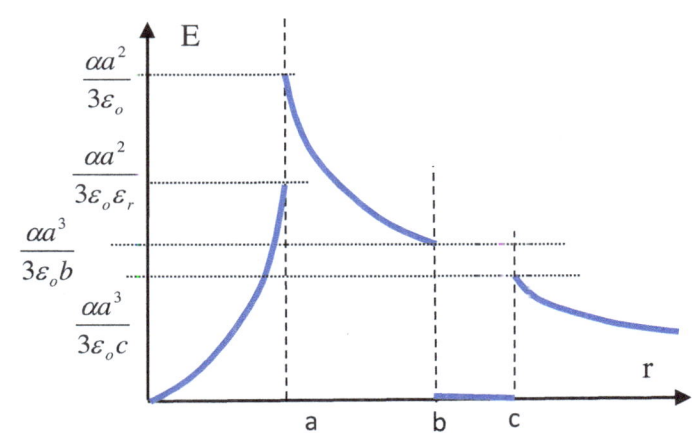

There is always a discontinuity in electric field at the boundary of a dielectric. It can be understood by examining the polarized charges. Supposed the total free charges enclosed is +Q at r = a, the total charge right below the surface would be less because the first layer of polarization charges are negative charges. Mathematically, it is because the dielectric constants are different inside or outside of the dielectric material.

Readers should be comfortable with setting up integrals, picking out typical elements, and more importantly, be able to visualize the integrating elements. Otherwise, please review applications of integral before trying out the problems at the end of the chapter.

2.10 Potential energy

Similar to the gravitational force, Coulomb force is a conservative force, which means there is a potential energy associates with Coulomb force and it only depends on locations. Equivalently, the closed loop integral is zero.

$$\oint \vec{F} \cdot d\vec{r} = 0 \qquad \text{Eqn.(2.13)}$$

Gravitational potential energy only depends on the height of the object. Likewise, electrical potential energy only depends on the position of the charged object relative to the other charges.

Stokes' Theorem stated that for any vector F:

$$\oint \vec{F} \cdot d\vec{r} = \int_S \nabla \times \vec{F} \cdot d\vec{a} \qquad \text{Eqn.(2.14)}$$

where S is any smooth surface bounded by the closed loop. For a conservative force, the left side of the equation is zero for any contour, so the integrand on the right side of Eqn.(2.14) must be zero. Since curl of a gradient is zero (Section 1.6.5), the conservative force is written as a gradient of a scalar function, which is called the potential energy.

$$\vec{F} \equiv -\nabla U,$$
$$U = -\int \vec{F} \cdot d\vec{r}, \qquad \text{Eqn.(2.15)}$$
$$\nabla \times \vec{F} = -\nabla \times \nabla U = 0.$$

The negative sign is introduced here so the definition of potential is consistent with the convention that objects always go from a higher potential to a lower potential location.

The gravitational force is very similar to the Coulomb force, so are the potential energies. The electrical energy stored between 2 point charges is:

$$U = k \frac{q_1 q_2}{r} \qquad \text{Eqn.(2.16)}$$

Since the charges can be positive or negative, a negative potential indicates attraction between charges, and a positive potential energy means repulsive. As usual, potential energy is also the work done required to put up such configuration.

Example 2.11: What is the potential energy stored in the 3-charge configuration in the diagram?

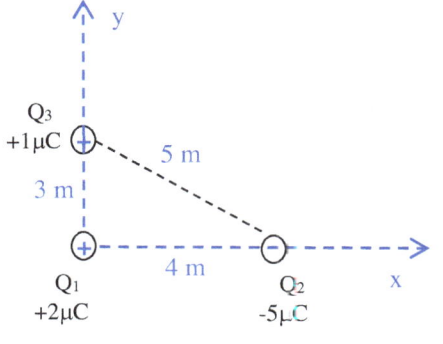

There are 2 ways to solve this problem:

1. Find the work required to put this configuration together. Let's bring in one charge at a time.
   To put the first charge (say +2µC) at the origin does not require any work since there is no other charges around.
   To bring in the 2nd charge (say -5µC) to the location (4,0) m would require:

$$U_{12} = k\frac{q_1 q_2}{r} = k\frac{(2)(-5)}{(4)} = -2.5k[10^{-2} J]$$

Negative work to bring in the -5µC means the system gains energy (it could be kinetic energy, for example). Now, bring in the Q$_3$ to the location (0, 3) m from infinity:

$$U_3 = U_{13} + U_{23} = k\frac{(1)(2)}{3} + k\frac{(1)(-5)}{5} = -\frac{k}{3}[10^{-12} J]$$

$$U_{total} = U_{12} + U_3 = \left(-2.5k - \frac{k}{3}\right)[10^{-12} J] = -2.83k[10^{-12} J] = -25.5[mJ]$$

2. Find half the sum of the potential energy of each charge. Half because the sum would have double-count the work to put the system together.

$$U_3 = U_{13} + U_{23} = k\frac{(1)(2)}{3} + k\frac{(1)(-5)}{5} = -\frac{k}{3}[10^{-12} J]$$

$$U_2 = U_{12} + U_{23} = k\frac{(2)(-5)}{4} + k\frac{(1)(-5)}{5} = -3.5k[10^{-12} J]$$

$$U_1 = U_{12} + U_{13} = k\frac{(2)(-5)}{4} + k\frac{(1)(2)}{3} = -\frac{11}{6}k[10^{-12} J]$$

$$U_{total} = \frac{1}{2}\sum_i U_i = \frac{1}{2}\left(-\frac{k}{3} - 3.5k - \frac{11}{6}k\right)[10^{-12} J] = -\frac{5.66}{2}k[10^{-12} J] = -25.5[mJ]$$

The ½ in Method #2 is the same ½ in the more familiar calculation of energy stored in a capacitor. U_{cap} = ½ CV^2 for the same reason.

It is easier to add scalars than vectors. Very often, people work with energy, and take the gradient of the total energy to get to the net force, as described in Eqn.(2.15).

2.11 Electric Potential

Electric potential is related to the potential energy as electric field is to the Coulomb force.

$$V \equiv \frac{U}{q} \qquad \text{Eqn.(2.17)}$$

$$\vec{E} = -\nabla V. \qquad \text{Eqn.(2.18)}$$

The difference in electric potential between 2 points in space is called the potential difference, and more often, referred to as the voltage between the 2 points. Similar to the electric field for a single point charge in Eqn.(2.4), one can define the electric potential for a point charge Q as:

$$V \equiv k\frac{Q}{r} \qquad \text{Eqn.(2.19)}$$

It is important to understand the physical meaning of U, V, E, F, and how they are related. The diagram below provides a useful roadmap, follows by a few examples of how to navigate between them.

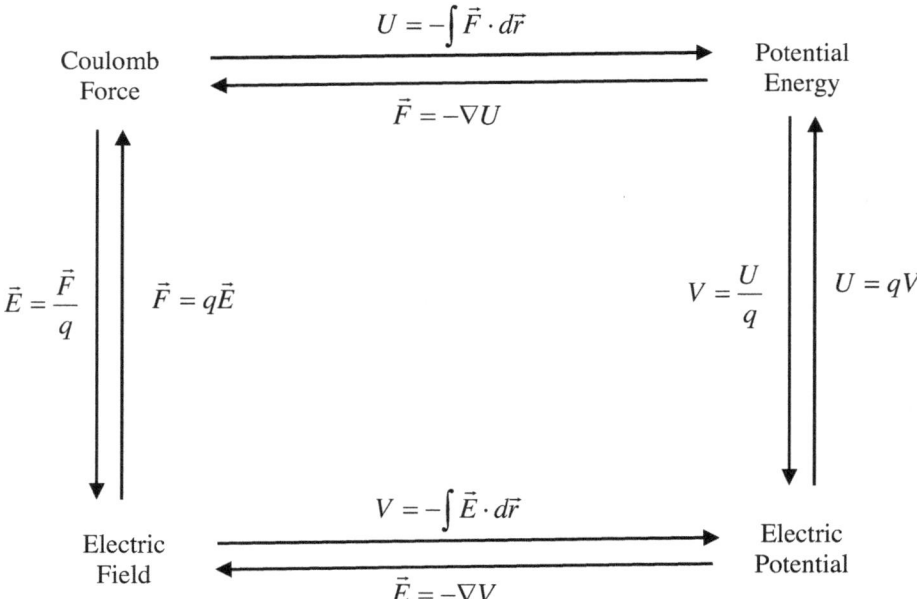

Fig.(2.13) – Relation between F, E, U and V.

Example 2.12: Find the electric potential and electric field of a finite line charge of length L, and a uniform linear charge density λ [in coul/m], at a distance x away from the center of the line.

Let's explore a couple of ways to solve this problem.

Method 1: Direct integral to get E, then calculate V from E.

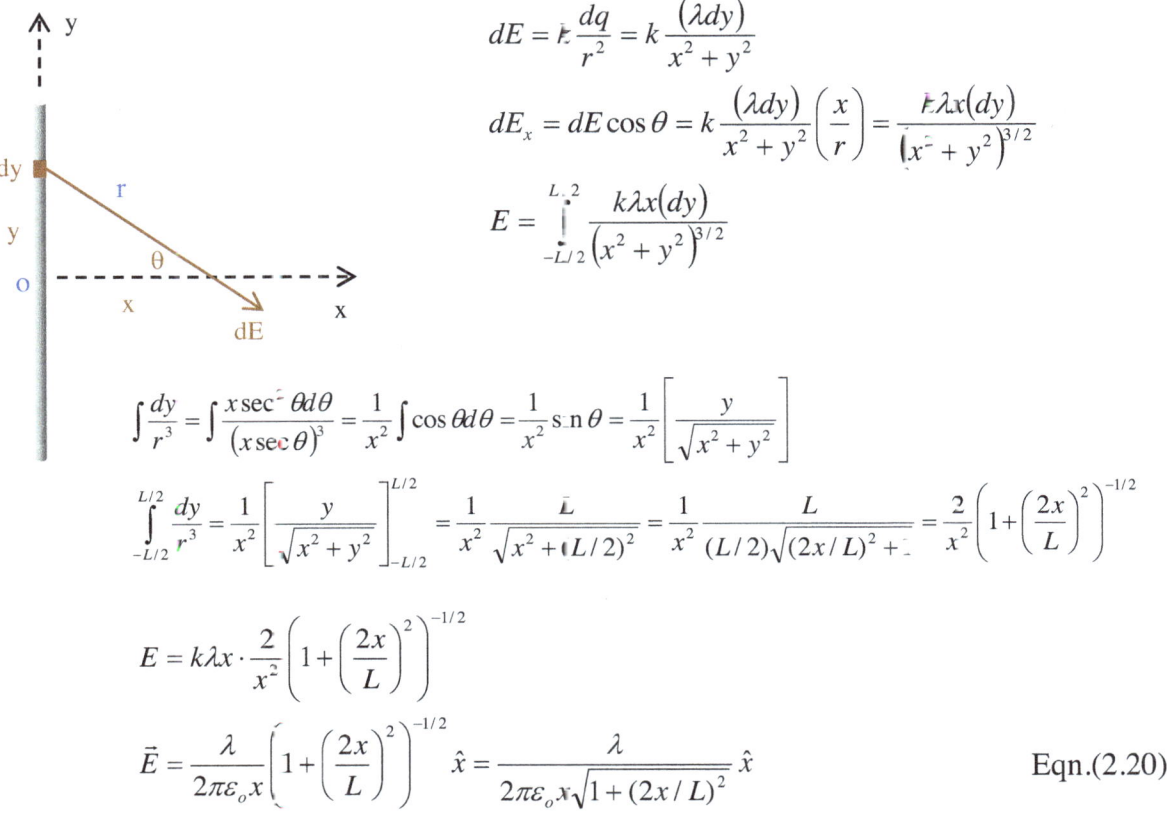

$$dE = k\frac{dq}{r^2} = k\frac{(\lambda dy)}{x^2 + y^2}$$

$$dE_x = dE \cos\theta = k\frac{(\lambda dy)}{x^2 + y^2}\left(\frac{x}{r}\right) = \frac{k\lambda x(dy)}{(x^2 + y^2)^{3/2}}$$

$$E = \int_{-L/2}^{L/2} \frac{k\lambda x(dy)}{(x^2 + y^2)^{3/2}}$$

$$\int \frac{dy}{r^3} = \int \frac{x\sec^2\theta d\theta}{(x\sec\theta)^3} = \frac{1}{x^2}\int \cos\theta d\theta = \frac{1}{x^2}\sin\theta = \frac{1}{x^2}\left[\frac{y}{\sqrt{x^2 + y^2}}\right]$$

$$\int_{-L/2}^{L/2} \frac{dy}{r^3} = \frac{1}{x^2}\left[\frac{y}{\sqrt{x^2 + y^2}}\right]_{-L/2}^{L/2} = \frac{1}{x^2}\frac{L}{\sqrt{x^2 + (L/2)^2}} = \frac{1}{x^2}\frac{L}{(L/2)\sqrt{(2x/L)^2 + 1}} = \frac{2}{x^2}\left(1 + \left(\frac{2x}{L}\right)^2\right)^{-1/2}$$

$$E = k\lambda x \cdot \frac{2}{x^2}\left(1 + \left(\frac{2x}{L}\right)^2\right)^{-1/2}$$

$$\vec{E} = \frac{\lambda}{2\pi\varepsilon_o x}\left(1 + \left(\frac{2x}{L}\right)^2\right)^{-1/2}\hat{x} = \frac{\lambda}{2\pi\varepsilon_o x\sqrt{1 + (2x/L)^2}}\hat{x} \qquad \text{Eqn.(2.20)}$$

Is this answer consistent with earlier result?

Recall the binomial expansion: $\underset{u \to 0}{Lim}(1+u)^n \approx 1 + nu$

Let $L \to \infty$, $\vec{E} = \underset{L \to \infty}{Lim}\frac{\lambda}{2\pi\varepsilon_o x}\left(1 + \left(\frac{2x}{L}\right)^2\right)^{-1/2}\hat{x} = \frac{\lambda}{2\pi\varepsilon_o x}\hat{x}$

which is the same as the infinite line charge result from Example 2.4.

Let $L \to 0$, $\vec{E} = \underset{L \to 0}{Lim}\frac{\lambda}{2\pi\varepsilon_o x\left(\frac{2x}{L}\sqrt{(L/2x)^2 + 1}\right)}\hat{x} = \frac{\lambda L}{4\pi\varepsilon_o x^2}\hat{x} = \frac{Q}{4\pi\varepsilon_o x^2}\hat{x}$

This is the point charge equation Eqn.(2.4).

33

For a finite line charge, V = 0 at ∞. This provides a convenient limit for the integral.

$$\int_0^V dV = -\int_\infty^x \vec{E} \cdot d\vec{\ell}$$

$$V = -\int_\infty^x \frac{\lambda}{2\pi\varepsilon_o x\sqrt{1+(2x/L)^2}} dx = -\frac{\lambda}{2\pi\varepsilon_o} \int_\infty^u \frac{du}{u\sqrt{1+u^2}} \qquad \text{where } u = 2x/L.$$

Trig substitution?

$$\int_\infty^u \frac{du}{u\sqrt{1+u^2}} = ?$$

$u = \tan\theta$
$du = \sec^2\theta\, d\theta$
$\sqrt{1+u^2} = \sec\theta$

$$\int \frac{du}{u\sqrt{1+u^2}} = \int \frac{\sec^2\theta\, d\theta}{\tan\theta \sec\theta} = \int \frac{d\theta}{\sin\theta} = \int \csc\theta\, d\theta = -\ln|\csc\theta + \cot\theta|$$

$$\int_\infty^u \frac{du}{u\sqrt{1+u^2}} = -\ln\left|\frac{\sqrt{1+u^2}}{u} + \frac{1}{u}\right|_\infty^u = \ln\left|\frac{u}{1+\sqrt{1+u^2}}\right|_\infty^u = \ln\left|\frac{u}{1+\sqrt{1+u^2}}\right|$$

U substitution?

$w^2 = 1+u^2$
$w\, dw = u\, du$

$$\int_\infty^u \frac{du}{u\sqrt{1+u^2}} = \int_\infty^u \frac{u\, du}{u^2\sqrt{1+u^2}} = \int \frac{w\, dw}{(w^2-1)w} = \int \frac{dw}{(w+1)(w-1)}$$

$$= \frac{1}{2}\int dw \left(\frac{1}{w+1} - \frac{1}{w-1}\right) = \frac{1}{2}\ln\left|\frac{w+1}{w-1}\right| = \frac{1}{2}\ln\left|\frac{\sqrt{1+u^2}+1}{\sqrt{1+u^2}-1}\right|$$

Are they not the same?

$$2\ln\left|\frac{1+\sqrt{1+u^2}}{u}\right| \Leftrightarrow \ln\left|\frac{\sqrt{u^2+1}+1}{\sqrt{u^2+1}-1}\right| \qquad ??$$

$$2\ln\left|\frac{1+\sqrt{1+u^2}}{u}\right| - \ln\left|\frac{\sqrt{u^2+1}+1}{\sqrt{u^2+1}-1}\right| = \ln\left|\left(\frac{1+\sqrt{1+u^2}}{u}\right)^2 \cdot \left(\frac{\sqrt{u^2+1}-1}{\sqrt{u^2+1}+1}\right)\right|$$

$$= \ln\left|\frac{1+\sqrt{1+u^2}}{u^2} \cdot \left(\sqrt{u^2+1}-1\right)\right| = \ln\left|\frac{1+u^2-1}{u^2}\right| = \ln|1| = 0 \qquad \text{Yes, they are the same.}$$

$$V = -\frac{\lambda}{2\pi\varepsilon_o}\ln\left|\frac{u}{\sqrt{1+u^2}+1}\right| = \frac{\lambda}{2\pi\varepsilon_o}\ln\left|\frac{\sqrt{1+(2x/L)^2}+1}{2x/L}\right| \quad, \text{ or}$$

$$V = -\frac{\lambda}{4\pi\varepsilon_o}\ln\left|\frac{\sqrt{1+u^2}+1}{\sqrt{1+u^2}-1}\right| = \frac{\lambda}{4\pi\varepsilon_o}\ln\left|\frac{\sqrt{1+(2x/L)^2}+1}{\sqrt{1+(2x/L)^2}-1}\right| \quad. \text{ Both answers are equally valid.}$$

Method 2: Direct integral to get V, then calculate E from V.

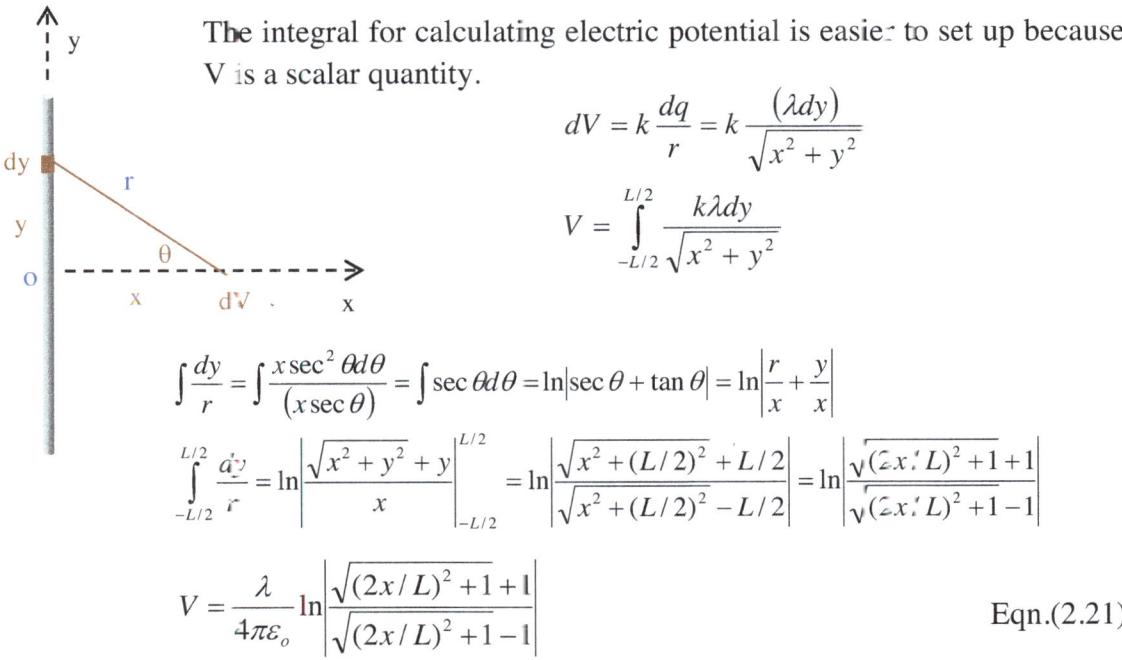

The integral for calculating electric potential is easier to set up because V is a scalar quantity.

$$dV = k\frac{dq}{r} = k\frac{(\lambda dy)}{\sqrt{x^2+y^2}}$$

$$V = \int_{-L/2}^{L/2} \frac{k\lambda dy}{\sqrt{x^2+y^2}}$$

$$\int \frac{dy}{r} = \int \frac{x\sec^2\theta d\theta}{(x\sec\theta)} = \int \sec\theta d\theta = \ln|\sec\theta + \tan\theta| = \ln\left|\frac{r}{x} + \frac{y}{x}\right|$$

$$\int_{-L/2}^{L/2} \frac{dy}{r} = \ln\left|\frac{\sqrt{x^2+y^2}+y}{x}\right|_{-L/2}^{L/2} = \ln\left|\frac{\sqrt{x^2+(L/2)^2}+L/2}{\sqrt{x^2+(L/2)^2}-L/2}\right| = \ln\left|\frac{\sqrt{(2x/L)^2+1}+1}{\sqrt{(2x/L)^2+1}-1}\right|$$

$$V = \frac{\lambda}{4\pi\varepsilon_o}\ln\left|\frac{\sqrt{(2x/L)^2+1}+1}{\sqrt{(2x/L)^2+1}-1}\right| \qquad \text{Eqn.(2.21)}$$

Again, let's check the point charge and infinite line charge limits.

$$V = \frac{\lambda}{4\pi\varepsilon_o}\ln\left|\frac{\sqrt{(2x/L)^2+1}+1}{\sqrt{(2x/L)^2+1}-1}\right| = \frac{\lambda}{4\pi\varepsilon_o}\ln\left|\frac{\sqrt{1+(L/2x)^2}+(L/2x)}{\sqrt{1+(L/2x)^2}-(L/2x)}\right| \approx \frac{\lambda}{4\pi\varepsilon_o}\ln\left|\frac{1+(L/2x)}{1-(L/2x)}\right|$$

$$\ln|1+x| = x - \frac{x^2}{2} + \frac{x^3}{3} - \frac{x^4}{4} + ... \approx x$$

$$\ln\left|\frac{1+(L/2x)}{1-(L/2x)}\right| = \ln\left|1+\frac{2(L/2x)}{1-(L/2x)}\right| \approx \frac{2(L/2x)}{1-(L/2x)} = \frac{2L}{2x-L} \approx \frac{L}{x}$$

$$V = \frac{\lambda}{4\pi\varepsilon_o}\left(\frac{L}{x}\right) = \frac{Q}{4\pi\varepsilon_o x} \qquad \text{Point charge result, as in Eqn.(2.19).}$$

And for the infinite line charge: L →∞,

$$\lim_{L\to\infty}\sqrt{(2x/L)^2+1} = \lim_{L\to\infty}\left(1+(2x/L)^2\right)^{1/2} \approx 1 + 2\left(\frac{x}{L}\right)^2$$

$$\ln\left|\frac{\sqrt{(2x/L)^2+1}+1}{\sqrt{(2x/L)^2+1}-1}\right| \approx \ln\left|\frac{2}{2(x/L)^2}\right| = 2\ln\left|\frac{L}{x}\right| = -2\ln|x| + C$$

$$V = -\frac{\lambda}{2\pi\varepsilon_o}\ln|x| + C$$

This is the infinite line charge electric potential, even though the constant is huge.

The infinite electric potential can be checked easily from the electric field $\vec{E} = \dfrac{\lambda}{2\pi\varepsilon_o x}\hat{x}$

$$V = -\int \vec{E}\cdot d\vec{\ell} = -\dfrac{\lambda}{2\pi\varepsilon_o}\int \dfrac{dx}{x} = -\dfrac{\lambda}{2\pi\varepsilon_o}\ln|x| + C$$

Electric field can be calculated by taking the gradient of V as stated in Eqn.(2.18).

$$V = \dfrac{\lambda}{4\pi\varepsilon_o}\ln\left|\dfrac{\sqrt{(2x/L)^2+1}+1}{\sqrt{(2x/L)^2+1}-1}\right| = \dfrac{\lambda}{4\pi\varepsilon_o}\left[\ln|\sqrt{u^2+1}+1| - \ln|\sqrt{u^2+1}-1|\right]$$

$$\vec{E} = -\hat{x}\dfrac{dV}{dx} = -\hat{x}\dfrac{2}{L}\dfrac{dV}{du} = -\hat{x}\dfrac{\lambda}{2\pi\varepsilon_o L}\left[\dfrac{\tfrac{1}{2}(u^2+1)^{-1/2}(2u)}{\sqrt{u^2+1}+1} - \dfrac{\tfrac{1}{2}(u^2+1)^{-1/2}(2u)}{\sqrt{u^2+1}-1}\right]$$

$$\vec{E} = -\hat{x}\dfrac{\lambda}{2\pi\varepsilon_o L}\dfrac{u}{\sqrt{u^2+1}}\left[\dfrac{(\sqrt{u^2+1}-1)-(\sqrt{u^2+1}+1)}{u^2+1-1}\right] = \hat{x}\dfrac{2\lambda}{2\pi\varepsilon_o Lu\sqrt{u^2+1}}$$

$$\vec{E} = \hat{x}\dfrac{\lambda}{2\pi\varepsilon_o x\sqrt{1+(2x/L)^2}}$$

In this simple example, the benefit of the second method is not apparent. But in general, it is much easier to calculate electric field from the scalar potential function. There is only one integral in V to evaluate, whereas the electric field has 3 integrals, one for each component. Moreover, differentiation is easier to execute than integration. Therefore, most commercial electromagnetic solvers take advantage of the potential functions.

Example 2.13: A thick spherical dielectric shell (ε_r = 3), with inner radius 2 m and outer radius 4 m, is filled with a charge density ρ = 5 r. Inner sphere (r < 2) is in vacuum.
(a) Find **E**(r) everywhere.
(b) Sketch E(r) and label all the points.
(c) What is the capacitance between the inner and outer radius?

(a) Gauss's Law:
For r < 2, E = 0 because there is no charge inside.

For 2 < r < 4,

$$\varepsilon EA = Q_{\substack{inside\\free}} = \int \rho dV = \int_2^r (5r)4\pi r^2 dr = 5\pi r^4 \Big|_2^r = 5\pi(r^4-16)$$

$$\varepsilon_r \varepsilon_o EA = (3)\varepsilon_o E 4\pi r^2 = 5\pi(r^4-16)$$

$$\vec{E}(2<r<4) = \dfrac{5}{12\varepsilon_o}\left(r^2 - \dfrac{16}{r^2}\right)\hat{r}$$

For $r > 4$, $\quad Q_{inside \atop free} = \int \rho dV = \int_2^4 (5r) 4\pi r^2 dr = 5\pi r^4 \Big|_2^4 = 5\pi(4^4 - 16) = 1200\pi$

$$E(4\pi r^2) = \frac{1200\pi}{\varepsilon_o}$$

$$\vec{E}(r > 4) = \frac{300}{\varepsilon_o r^2} \hat{r}$$

(b)

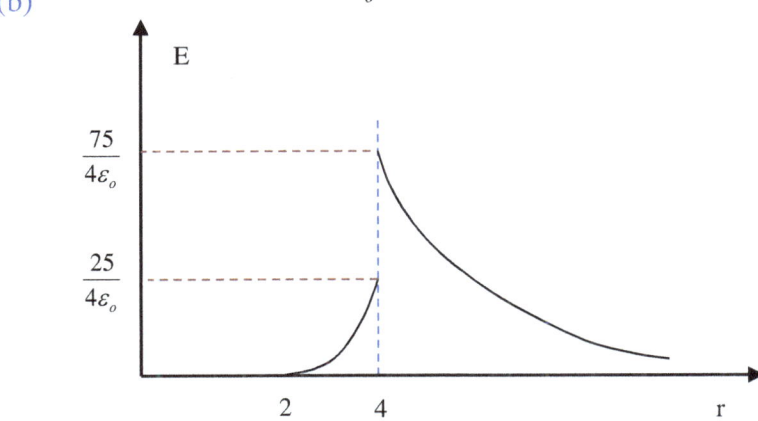

(c)

$$V = -\int_4^2 \vec{E} \cdot d\vec{r} = \frac{5}{12\varepsilon_o} \int_2^4 \left(r^2 - \frac{16}{r^2} \right) dr = \frac{5}{12\varepsilon_o} \left[\frac{r^3}{3} + \frac{16}{r} \right]_2^4 = \frac{55}{9\varepsilon_o}$$

$$C = \frac{Q}{V} = \frac{1200\pi}{55/9\varepsilon_o} = 196\pi\varepsilon_o = 5.5 nF$$

2.12 Image charge technique

Fig.(2.14) – Seeing a reflection off a mirror.

A ground plane in electromagnetism is similar to a mirror in optics. They both create images. But what is an image? It is a simplified description of how the light reflects off the mirror surface, as if the light was coming from behind the mirror. In other words, looking in front of a perfect mirror, there is no way to distinguish between 2 objects without the presence of a mirror, or an object in front of the mirror.

People usually can tell whether there is a mirror presence, by checking the surrounding and familiar objects such as a door or a table top, by looking for the edge of the mirror, or by seeing the imperfection on the mirror surface. Our life experience comes in rescue in these cases. However, there are plenty of funny videos on small pets seeing their images the first time. Without looking at the base of the vase in Fig.(2.14), there is no way to tell one of the rose is an image behind the mirror.

2.12.1 A point charge in front of an infinite ground plane.

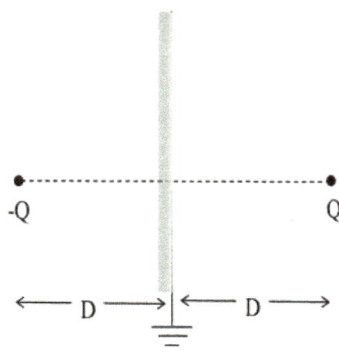

Fig.(2.15) – A single charge in front of a ground plane.

A ground plane is an equipotential conducting surface, and the constant electric potential is taken as zero. When a charge +Q is brought close to the plane, negative charges are induced on the surface. The induced surface charge density would depend on the location on the plane, and would not be uniform. Without knowing the charge density, the integration technique would not be possible.

To replace the flat, infinite ground plane, one simply put a negative image charge at the same distance behind the plane. This is to guarantee the zero potential on the plane location. With the image charge, calculate electric field in front of the plane becomes summing the electric field of 2 point charges.

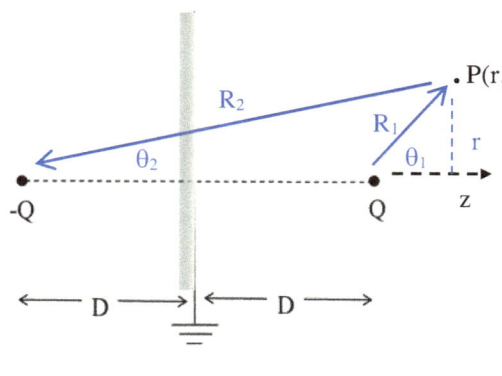

Because of the symmetry about the z-axis, using cylindrical coordinates to find the electric field in front of the plane would be more natural and easier.

$$\vec{E}_1 = k\frac{Q}{R_1^2}(\sin\theta_1, 0, \cos\theta_1) = \frac{Q}{4\pi\varepsilon_o}\frac{[\hat{r}r + \hat{z}(z-D)]}{[r^2 + (z-D)^2]^{3/2}}$$

$$\vec{E}_2 = k\frac{(-Q)}{R_2^2}(\sin\theta_2, 0, \cos\theta_2) = \frac{-Q}{4\pi\varepsilon_o}\frac{[\hat{r}r + \hat{z}(z+D)]}{[r^2 + (z+D)^2]^{3/2}}$$

$$\vec{E}_{total} = \frac{Q}{4\pi\varepsilon_o}\left\{\frac{[\hat{r}r + \hat{z}(z-D)]}{[r^2 + (z-D)^2]^{3/2}} - \frac{[\hat{r}r + \hat{z}(z+D)]}{[r^2 + (z+D)^2]^{3/2}}\right\}$$

With this result, one can calculate the surface charge density on the ground plane. This will be addressed later in the chapter of Maxwell Equations.

2.12.2 A point charge in front of an infinite ground plane corner.

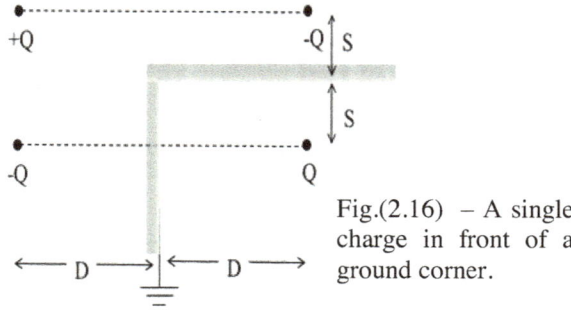

Fig.(2.16) – A single charge in front of a ground corner.

When a person stands in front of a corner with 2 mirrors perpendicular to each other, multiple images are produced. The same is true for a charge placed in front of 2 ground planes joined at 90 degrees, multiple image charges appear.

In Fig.(2.16), one can think of removing the vertical ground plane first, and replace it with an image charge at –D. But at the same time, the horizontal ground plane should also be extended to the -∞ direction. Now, removing the horizontal ground plane (and its extension) would require 2 more image charges as shown.

What happens if the 2 mirrors are joined at an angle of 45°? How many image charges should there be? It would be a good idea to play with 2 mirrors at various angles, and observe the images they produce.

2.12.3 A point charge in front of a ground sphere

In optics, an object in front of a convex mirror would produce a smaller virtual image behind the mirror. In electrostatic, a point charge placed in front of a grounded sphere would produce a weaker image charge somewhere behind the spherical surface.

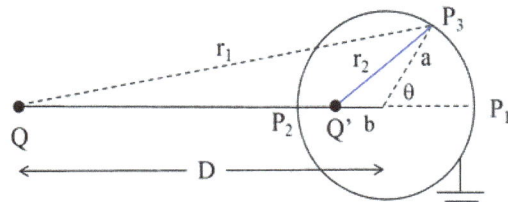

Fig.(2.17) – A point charge in front of a grounded sphere.

Again, the grounded surface is to provide an equipotential surface of V = 0

In Fig.(217), Q' is the image charge located at a distance "b" away from the center of the sphere of radius "a". The task is to evaluate the strength of the image charge, and locate it.

At point P_1, $V_1 = k\dfrac{Q}{D+a} + k\dfrac{Q'}{a+b} = 0$ $\Rightarrow Q(a+b) = -Q'(D+a)$

At point P_2, $V_2 = k\dfrac{Q}{D-a} + k\dfrac{Q'}{a-b} = 0$ $\Rightarrow Q(a-b) = -Q'(D-a)$

Combining these 2 equations gives: $b = \dfrac{a^2}{D}$, and $Q' = -\dfrac{a}{D}Q$.

In short, by placing an image charge Q' at the location "b" as shown, the spherical equipotential surface is maintained at V = 0. This can be checked at any point of the surface, say P_3:

$V_3 = k\dfrac{Q}{r_1} + k\dfrac{Q'}{r_2}$ where $\begin{aligned} r_1 &= \sqrt{D^2 + a^2 + 2Da\cos\theta} \\ r_2 &= \sqrt{b^2 + a^2 + 2ba\cos\theta} \end{aligned}$

Note: $\dfrac{Q'}{r_2} = \dfrac{-\dfrac{a}{D}Q}{\sqrt{\left(\dfrac{a^2}{D}\right)^2 + a^2 + 2\left(\dfrac{a^2}{D}\right)a\cos\theta}} = \dfrac{-Q}{\sqrt{a^2 + D^2 + 2Da\cos\theta}} = -\dfrac{Q}{r_1}$. So $V_3 = 0$. ✓

2.12.4 A charged conducting sphere in front of an infinite ground plane

Finite charged objects is a lot more complicated than point charges. Here is an example of problems dealing with more realistic geometries.

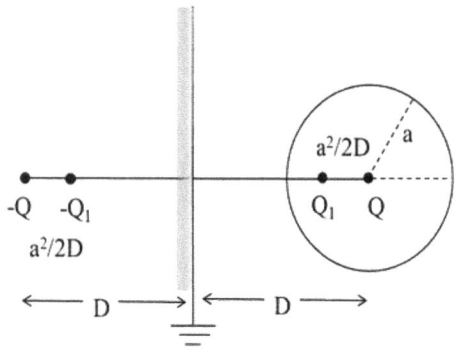

A conducting sphere with total charge Q on the surface is placed in front of an infinite ground plane as shown. Identify the image charges.

To find the electric field outside of the sphere, we can treat all the charges on the sphere locate at the center, as illustrated in Example (2.9). That would make no different because the electric potential on the surface would be equal.

Now the problem reduced to a point charge in front of an infinite ground plane. This would induce an image charge of strength –Q at a distance D behind the ground plane. The charge pair Q and Q' guarantees V = 0 on the ground plane. But that screws up the equipotential surface on the _{conductiing} sphere. To correct that, a second image charge Q₁ is introduced according to the formulas from the last section. Namely,

$$Q_1 = -\frac{a}{(2D)}(-Q) = \frac{a}{2D}Q \qquad \text{at} \qquad b_1 = \frac{a^2}{(2D)} \quad .$$

This pair of charges Q' and Q₁ provides an equipotential surface for the sphere. But now, it offset the equipotential requirement on the ground plane. To fix this, the 3rd image with strength –Q₁ appears at a distance (D – b₁) behind the plane. By now, it is clear this is going to be an infinite series of image charges, each successive image with decreasing strength. The strength of the first few images is summarized below. Readers should try to verify this. The expression gets complicated very fast. But the strength of the image charges also drop rapidly for a << D.

$$u \equiv \frac{a}{2D}$$

$$Q_1 = uQ$$

$$Q_2 = \frac{u^2}{1-u^2}Q$$

$$Q_3 = \frac{u^3}{(1-u^2)\left(1-\dfrac{u^2}{1-u^2}\right)}Q$$

In reality, the series converge rapidly and there is no need to carry out too many terms in the calculation. With the modern computer, this computation becomes very manageable.

Exercise:

2.1 Consider two point charges (+Q, -Q) located along the x-axis as shown.
   a. What is the electric potential at point P(2a,0)? (Note: there is no charge at point P.)
   b. From this potential, derive an equation for the electric field at that point P.
   c. How much work is required to bring another point charge (+q) from infinity to this point P? What does the sign mean?
   d. How much energy is stored in this system of 3-charges? (+Q, -Q, +q).

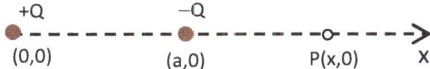

2.2 A conducting cylinder of infinite length and radius "a" has a total charge Q uniformly distributed on its surface. A dielectric shell (with relative dielectric constant $\varepsilon_r = 2$) of inner radius "b" and outer radius "c" is shielding the conducting rod as shown in the Figure. Calculate the electric field in all regions. Sketch **E**(r) as a function of r.

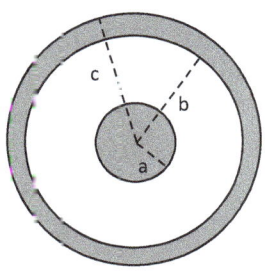

2.3 (a) What is the electric field at a distance x away from the center of a thin line charge with a uniform linear charge density λ and length L? (see diagram below). (b) Using this result, write the electric field at a distance x above the center of a square loop of length L.

(a)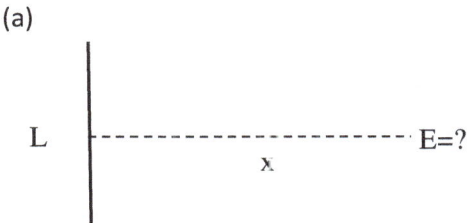

(b)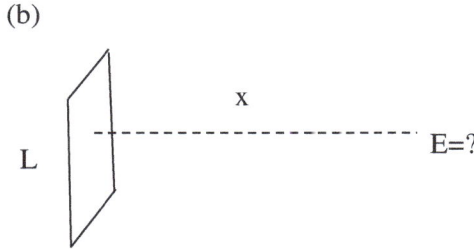

2.4 Consider 4 point-charges (+Q, +Q, -Q, -Q,) located at the corners of a square as shown.
   (a) List all the locations where electric field is zero.
   (b) List all the locations where electric potential is zero.

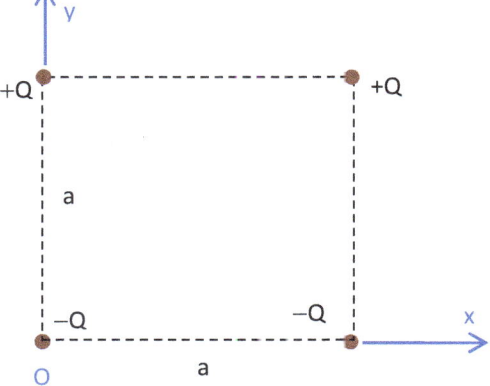

41

2.5 A non-uniform charge density of $\rho = \alpha r^2$ is distributed in an inner sphere of radius "a", enclosed by an outer spherical conductor with radius "b" and "c" as shown in diagram. The outer conductor is electrically grounded. Calculate the electric field in ALL regions. Sketch $\mathbf{E}(r)$ as a function of r. What is the voltage between the inner sphere and the outer conductor? What is the equivalent capacitance of the structure?

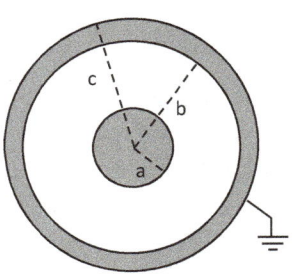

2.6 Consider two point charges (+Q, -Q) located at the base of an equilateral triangle as shown.
(a) What is the electric potential at point P (vertex)? (See diagram.)
(b) What is the electric field at point P?
(c) How much work is required to bring another point charge (+q) from infinity to this point P?
(d) How much energy is stored in this system of 3-charges? (+Q, -Q, +q).

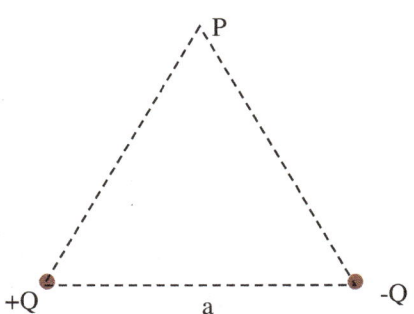

2.7 A coaxial cable is made of an inner cylindrical conductor with radius "a", an outer cylindrical conductor with radius "b" and "c" as shown in diagram. The outer conductor is electrically grounded. A dielectric material (ε_r) is filled between the 2 conductors (a < r < b). Suppose we have an infinite length of such cable, and a surface charge density σ is introduced onto the inner conductor, calculate the electric field in ALL regions. Sketch $\mathbf{E}(r)$. What would the voltage between the 2 conductors be? What is the equivalent capacitance per length (i.e. the total capacitance divided by the total length)?

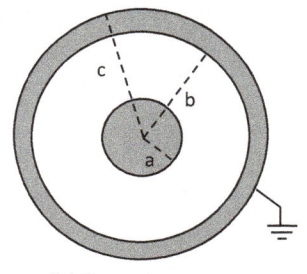

2.8 What is the electric field at a distance x away from the center of a thin line charge (along the z-axis) with length L and a non-uniform linear charge density $\lambda = 3z$?

2.9 What is the electric field at a distance x away from the center of a square plane with length L and a non-uniform surface charge density $\sigma = 3z$?

2.10 Find the electric field at a distance z above an infinite charged plate with surface charge density σ, by integrating point charges in Cartesian coordinates. See Section 2.6.3.

2.11 In Fig.(2.16), find the electric potential and electric field at a point P(0,y,z) in front of the grounded corner.

3. Magnetism

Early compass, called "south-pointing needle", was used in prehistoric Chinese battles around 2500 BC. Since then, magnet has been discovered and widely used around the globe. Applications include fortune telling and "feng shui" which has thousands of years history, and more recent computer memories in data storage technology. But not until the late 18th century, the quantitative magnetic force and field equations were formulated successfully.

3.1 Magnetic Force Equation

The most fundamental observable magnetic force equation is written as:
$$\vec{F} = q\vec{v} \times \vec{B} \qquad \text{Eqn.(3.1)}$$
where q is the electrical charge in coulombs, v is the velocity of the charge in m/s, and B is the magnetic field strength in teslas.

3.2 Comparison between the magnetic force $\vec{F}_B = q\vec{v} \times \vec{B}$, and the electrical force $\vec{F}_e = q\vec{E}$

(i) Velocity is not in the electrical force formula. So all charges in the presence of an electric field experience an electrical force. On the other hand, magnetic field only affects moving charges.

(ii) Magnetic force is a cross product. So the force is perpendicular to both the velocity and the magnetic field. A force perpendicular to the motion (centripetal force) would make the object goes in a circular motion (or spiral if the velocity also has a component along the B field). The electrical force, however, accelerates the charge in the same direction of the electric field.

(iii) Centripetal force does not increase kinetic energy of the object. But electric field does accelerate the charge and thus increases its kinetic energy.

3.3 Direction of magnetic force $\vec{F}_B = q\vec{v} \times \vec{B}$

As a cross product, basic right-hand-rule applies, as discussed in Chapter 1.

Example 3.1: What is the magnitude and direction of the magnetic force acting on a positive charge q, and describe the motion particle thereafter.

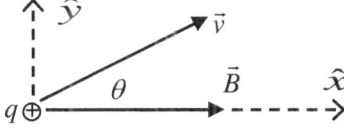

Magnitude of F = qvBsinθ, its direction is into the page (-z direction). Note: the coordinate systems are always defined by the right-hand-rule. With the directions of x and y axes given, z-axis is pointing out of the page. The charge q is going to spiral around the x-axis and move to the right.

3.4 In terms of current I

Current is basically moving charges. The fundamental macroscopic magnetic force can be rewritten as $\vec{F} = I\vec{L} \times \vec{B}$ Eqn.(3.2)
where L is the length of the wire that carries the current I. The direction of $\vec{L}$ is of course the direction of the current. Current is defined as $I = \frac{dq}{dt}$ and is not a vector itself, since q and t are both scalar quantities.

3.5 Ampere's Law

The origin of magnetic field was not fully understood until Einstein published his Theory of Special Relativity in 1905. But the idea that moving charges, or current, creates magnetic field was accidentally discovered in 1820 by the Denmark physicist Hans Orsted while he was preparing his demonstration on electric circuits the day before his lecture. Later that year, Ampere published his famous Ampere's Law which stated:

$$\oint \vec{B} \cdot d\vec{\ell} = \mu_o I_t$$ Eqn.(3.3)

where I_t is the total current passes through the closed loop in the integral, μ_o is the free space permeability equals to $4\pi \times 10^{-7}$ (or 1.2566×10^{-6}) in SI unit.

3.6 Ampere's Loop

Similar to the Gauss's Law in electrostatic, the integral in Eqn.(3.3) would be much simplified if one chooses a contour that has the same symmetry as the current distribution, such that:
(1) The magnitude of B is uniform on the contour.
(2) The direction of B lines up with the contour, so the dot product is always equal to 1.

Such a contour is sometimes referred to as an "Ampere's Loop". And the integral in Eqn.(3.3) is now simplified an algebraic equation:
$BL = \mu_o I_t$ Eqn.(3.4)

where L is the length of the contour. Again, only the current enclosed by this contour would contribute to the magnetic field.

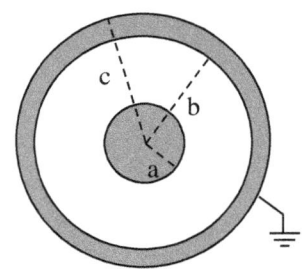

Example 3.2: An infinitely long coaxial cable with grounded shield is shown in figure. A non-uniform current density of J = αr (pointing out of the page, in z-direction) is distributed in an inner cylinder of radius "a". The outer conducting shield with inner and outer radius "b" and "c" is grounded at the exterior

surface as shown. The center cylinder is supported by Teflon with $\varepsilon_r = 2$ between "a" and "b". Assuming all the current returns through the surface of the outer conductor (at r = b). Calculate the magnetic field in ALL regions. Sketch B(r) as a function of r. What is the direction of this B-field? What is the induced surface current density at r = b?

For r < a

$$\oint \vec{B} \cdot d\vec{\ell} = \mu_o I_{inside}$$

$$B(2\pi r) = \mu_o \int_0^r \vec{J} \cdot d\vec{A} = \mu_o \int_0^r (\alpha r)(2\pi r dr)$$

$$B(2\pi r) = \mu_o 2\pi \alpha \left[\frac{r^3}{3} \right]$$

$$\vec{B} = \frac{\mu_o \alpha}{3} r^2 \hat{\phi}$$

For a < r < b

$$B(2\pi r) = \mu_o \int_0^a \vec{J} \cdot d\vec{A} = \mu_o 2\pi \alpha \left[\frac{a^3}{3} \right]$$

$$\vec{B} = \frac{\mu_o \alpha a^3}{3r} \hat{\phi}$$

B(b < r < c) = 0
(given, no current in shield, grounded)

B(r > c) = 0
(shielded)

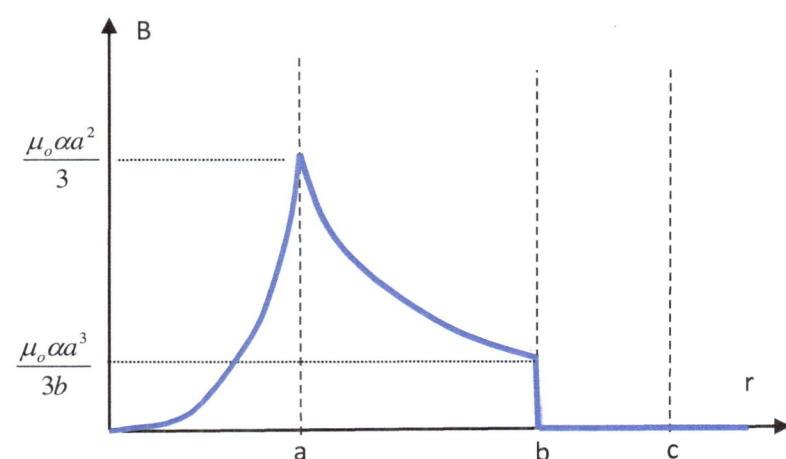

Total surface current on the inner surface of the grounded conductor has to be equal and opposite to the total current carried by the inner conductor, so that taking Ampere's Law around a contour in the shield would be zero.

$$\vec{J}_s(r=b) = \frac{-I_{total}}{length} = \frac{-2\pi \alpha \left[\frac{a^3}{3} \right]}{2\pi b} = \frac{-\alpha a^3}{3b} \hat{z}$$

Similar to the Gauss's Law, Ampere's Law is useful only if an Ampere's loop can be identified. There are only a few cases where this is true. Nonetheless, Ampere's Law is still a good first approximation in many situations.

3.7 Magnetic field in a metal

Magnetic behavior of metal is a topic of its own. There is no general rule to say that the magnetic field has to be zero, especially with direct current. Only in superconductor, B = 0 (known as Meissner effect) and that is not just because the resistance is zero. At higher frequency, the inability to penetrate a good conductor by both electric and magnetic field will be discussed in a later chapter.

Example 3.3: Magnetic force between 2 long parallel, current-carrying wires.

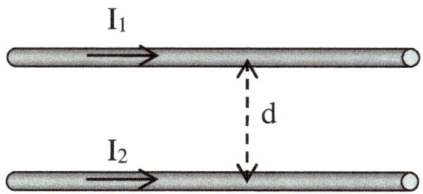

Two long wires carrying current I_1 and I_2, are separated by a distance d as shown.

The top current produces a magnetic field around the top wire according to Ampere's Law. The 2nd current in the bottom wire reacts to this magnetic field and experiences a magnetic force. What is the strength and direction of this force?

The magnetic field created by the current I_1 at the distance d away is: $\oint \vec{B}_1 \cdot d\vec{\ell} = \mu_o I_{inside}$

$$B_1(2\pi d) = \mu_o I_1$$

$$B_1 = \frac{\mu_o I_1}{2\pi d} \otimes$$

The B-field is pointing into the page. Now the wire #2 with current I_2 is experiencing a magnetic force due to the B-field created by I_1 :

$$\vec{F}_{21} = I_2 \vec{L}_2 \times \vec{B}_1 = I_2 L \cdot \frac{\mu_o I_1}{2\pi d} \uparrow$$

Likewise, one can calculate the magnetic field created by I_2 at a distance d: $B_2 = \frac{\mu_o I_2}{2\pi d} \odot$

and the magnetic force experienced by the wire #1 :

$$\vec{F}_{12} = I_1 \vec{L}_1 \times \vec{B}_2 = I_1 L \cdot \frac{\mu_o I_2}{2\pi d} \downarrow$$

Of course, $\vec{F}_{12} = -\vec{F}_{21}$ as predicted by Newton's Third Law of motion. The attraction force comes in pairs, and they are equal but opposite to each other.

Example 3.4: An ideal toroid.

A toroid is a doughnut-shaped toroidal solenoid as shown. In practice, it has many more turns of wire closely packed than it is in figure. Use Ampere's Law to find B field as a function of r (for N turns of wire carrying a current I). What is the direction of this B field? [The inner radius of the toroid is a, and the outer radius is b. Assuming the inner core is non-magnetic.]

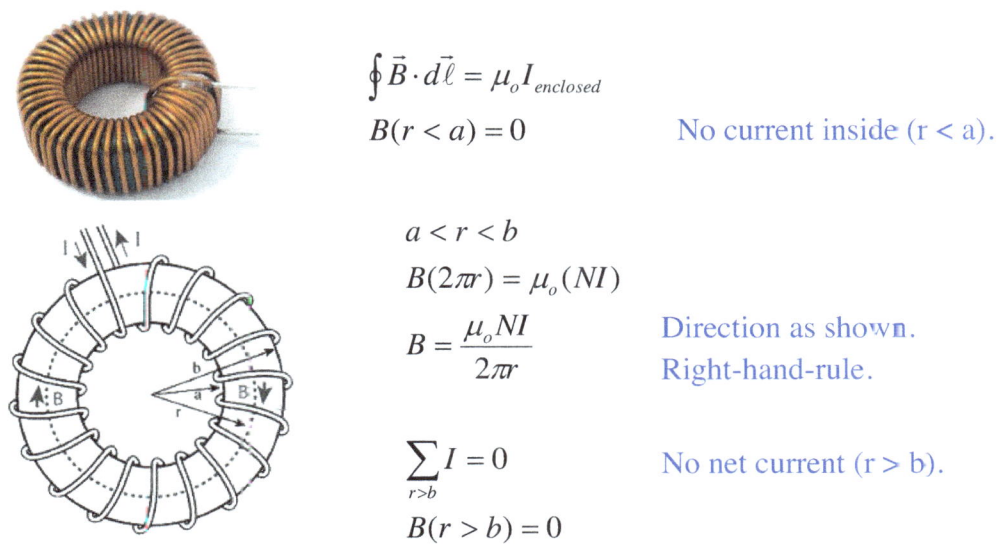

$$\oint \vec{B} \cdot d\vec{\ell} = \mu_o I_{enclosed}$$

$B(r < a) = 0$ 　　　No current inside (r < a).

$a < r < b$
$B(2\pi r) = \mu_o (NI)$
$B = \dfrac{\mu_o NI}{2\pi r}$ 　　　Direction as shown. Right-hand-rule.

$\sum_{r>b} I = 0$ 　　　No net current (r > b).
$B(r > b) = 0$

If the inner core is made of ferrite, we will have to use a modified Ampere's Law similar to the way that Gauss's Law was modified to take the polarization charges into account. The approximation is simple enough to understand with the introduction of the H-field later. However, in practice, the calculation is very complex and non-linear. I will leave this and other material related topics for future discussion.

3.8　Biot-Savart Law (The BS Law)

Electrical charges are the source of electric field, whereas electrical currents are the source of magnetic field. This was suggested by Orsted's discovery in 1820 as discussed in Section 3.5. Biot and Savart then derived a fundamental relationship of the magnetic field to a current element, known as the Biot-Savart Law:

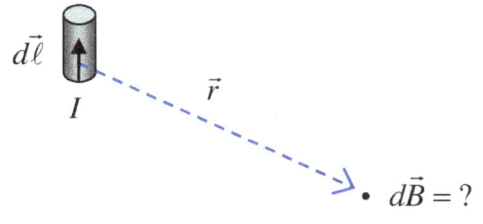

$$d\vec{B} = \dfrac{\mu_o}{4\pi} \dfrac{I d\vec{\ell} \times \hat{r}}{r^2} \qquad \text{Eqn.(3.5)}$$

where $\vec{r}$ is a distance vector defined from the current element to the point the B field is being calculated.

Example 3.5: Find magnetic field at a distance x away from an infinite thin wire carrying current I.

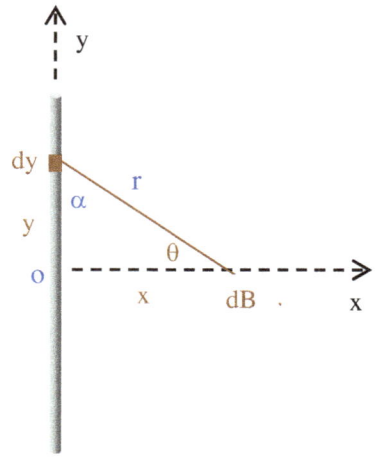

Using Ampere's Law:

$$\oint \vec{B} \cdot d\vec{\ell} = \mu_o I_{inside}$$

$$B(2\pi x) = \mu_o I$$

$$\vec{B} = \frac{\mu_o I}{2\pi x} \hat{\phi}$$

curves around the wire.

Using Biot-Savart Law: $d\vec{B} = \frac{\mu_o}{4\pi} \frac{I d\vec{\ell} \times \hat{r}}{r^2} = \frac{\mu_o}{4\pi} \frac{I dy \sin\alpha}{r^2} = \frac{\mu_o}{4\pi} \frac{I dy \cos\theta}{r^2} \otimes$

pointing into the page regardless where y is.

$$B = \frac{\mu_o I}{4\pi} \int_{y=-\infty}^{\infty} \frac{\cos\theta \, dy}{r^2} = \frac{\mu_o I}{4\pi} \int \frac{\cos\theta (x \sec^2\theta \, d\theta)}{(x \sec\theta)^2}$$

$$\vec{B} = \frac{\mu_o I}{4\pi x} \int_{-\pi/2}^{\pi/2} \cos\theta \, d\theta = \frac{\mu_o I}{4\pi x}[2] = \frac{\mu_o I}{2\pi x} \otimes \quad \text{same result}$$

Example 3.6: A circle of radius R with its center at the origin as shown is carrying a steady current of I. Calculate the magnitude AND direction of the magnetic field at (0,0,z) in this coordinate.

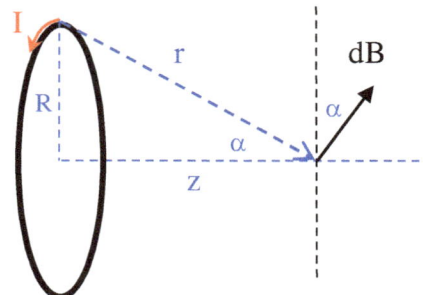

It is not easy to show the 3-dimensional nature on a 2-dimensional drawing. But being able to see the vector dB is the most crucial step of solving this problem.

From the symmetry of the geometry and the right-hand-rule, one can see that the net magnetic field at (0,0,z) is pointing to the right, along the z-direction. So, only the z-component is needed for the integral.

$$d\vec{B} = \frac{\mu_o}{4\pi} \frac{I d\vec{l} \times \hat{r}}{r^2} = \frac{\mu_o}{4\pi} \frac{IR d\theta \sin(90^o)}{r^2}$$

$$dB_z = dB \sin\alpha = \frac{\mu_o}{4\pi} \frac{IR d\theta}{r^2} \frac{R}{r} = \frac{\mu_o}{4\pi} \frac{IR^2 d\theta}{r^3}$$

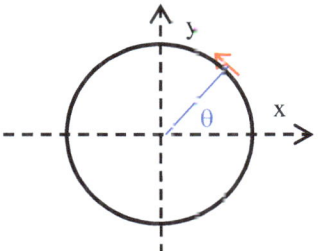

The angle θ is the integration variable, define by the location of the typical current element on the circle, as shown. Readers should have a solid understanding of calculus and be able to visualize it to avoid confusion between various angles in the problem.

$$\vec{B} = \frac{\mu_o IR^2}{4\pi r^3} \int_0^{2\pi} d\theta = \frac{\mu_o IR^2}{2(R^2 + z^2)^{3/2}} \hat{z}$$

A special case when $z = 0$, $B = \mu_o I / 2R$ is the magnetic field at the center of a current ring, and its direction is given by the right-hand-rule. The curving fingers represent the curving direction of the current, and the thumb, pointing straight up, represents the magnetic field at the center is pointing up.

3.9 Magnetic moment

When a current loop is placed in the presence of a magnetic field, the magnetic force exerts on the loop will cause the loop to rotate if it is pivoted about a rotational axis. DC motors make use of this simple but powerful phenomenon. Fig.(3.1) shows a simple dc motor made of paperclips, a wire loop, a battery and a magnet.

Fig.(3.1) – A simple DC motor.

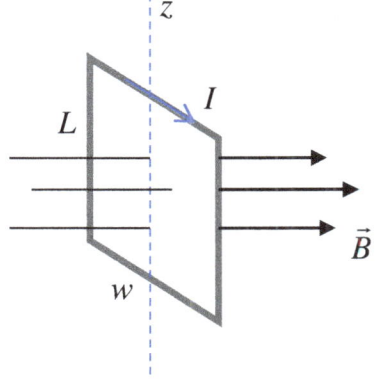

Fig.(3.2) – A rectangular loop submerged in a B-field.

Consider a rectangular loop of wire, with length L and width w as shown in figure, submerges in a constant magnetic field in the x-direction. Suppose the loop is pivoted about the z-axis and is free to rotate. What is the magnetic torque acting on this loop? Once the torque is calculated, other kinematic information can be extracted, such as angular velocity at a given time….etc.

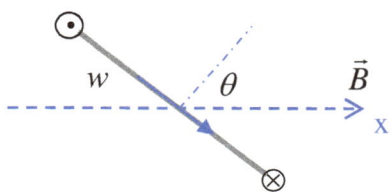

Fig.(3.2a) – Top view of Fig.(3.2)

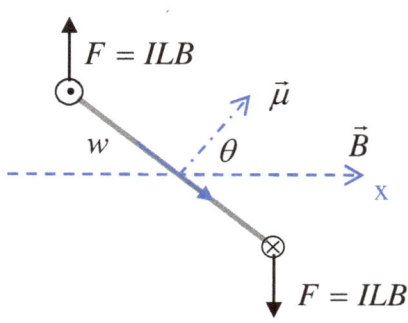

Fig.(3.2b) – Same as Fig.(3.2a), adding the force direction.

Perhaps it is easier to see the loop from the top view, depicted in Fig.(3.2a). Also shown are the direction of the current on the top wire, and the direction on the 2 sides. Again, the magnetic field is pointing along the x-axis.

On the lower edge of the loop, where the current is going into the page, the magnetic force is: $\vec{F} = I\vec{L} \times \vec{B} = ILB \downarrow$

On the upper edge where the current is coming out of the page, the force is: $\vec{F} = I\vec{L} \times \vec{B} = ILB \uparrow$

The force on the top wire is: $\vec{F} = I\vec{L} \times \vec{B} = IwB\hat{z}$

And the force on the bottom wire is: $\vec{F} = I\vec{L} \times \vec{B} = -IwB\hat{z}$

Clearly the net force acting on the loop is zero. However, the net torque is not. This is clearly illustrated in Fig.(3.2b). The net torque about the origin is:

$$\vec{\tau} = \sum \vec{r} \times \vec{F} = 2\left[-\hat{z}\left(\frac{w}{2}\right)F\sin\theta\right] = -\hat{z}wILB\sin\theta$$

In Fig.(3.2b), z-axis is pointing out of the page, so the torque is pointing into the page. It really means the loop is going to rotate (and accelerate) in the clockwise direction in the view of Fig.(3.2b) at that instance.

In mathematics, area can be represented as a vector, with magnitude equals to length times the width, and the direction is the normal to the surface. In Fig.(3.2b), it would be pointing along the dash line at an angle θ with a magnitude of Lw. The vector $\vec{\mu}$ shown is called the magnetic moment, defined by: $\vec{\mu} \cong I\vec{A}$. With this, the net torque on the loop can be written as:

$$\vec{\tau} = \vec{\mu} \times \vec{B}$$

Eqn.(3.6)

$$\vec{\tau} = I\vec{A} \times \vec{B} = -\hat{z}ILwB\sin\theta.$$

Basically, the magnetic moment μ describes the magnetic behavior of a current element. A Bohr atom is a prime example. With the electrons orbiting around the nucleus, an atom is equivalent to a current element in a circular loop. In other words, a simple atom behaves like a small magnet, and hence μ is also called the magnetic dipole moment. How do these dipole moments interact among themselves in a solid derive the magnetic properties of various solid. More importantly, Eqn.(3.6) describes the interaction between the object (an atom, a molecule, a nucleus, a rectangular loop of wire…etc) and the external magnetic field. Many of these interactions were further developed and used in instrumentations, such as MRI (magnetic resonance imaging), ESR (electron spin resonance), and μSR (muon spin rotation)…etc.

3.10 Image current

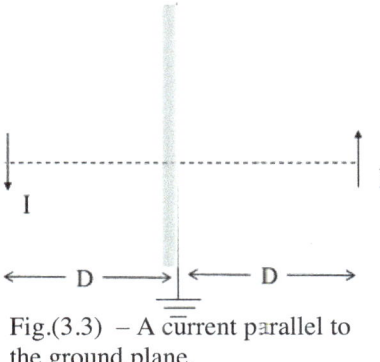

Fig.(3.3) – A current parallel to the ground plane.

When a current element is placed in front of a ground plane, an image current is produced behind the plane. The direction of the current depends on the orientation of the current. Again the concept of image charge and current is an effective equivalent model to calculate the electromagnetic field in front of the ground plane.

Fig. (3.3) shows a current parallel to the ground plane, produces an image current in the opposite direction.

Imagine a positive charge placed in front of the ground plane, a negative image charge is induced behind the plane. If the positive charge is moving upward, effectively producing a current going up, the image negative charge would follow and move upward. But negative moving upward means a positive current is moving downward. Therefore, in this case, the image current is in the opposite direction.

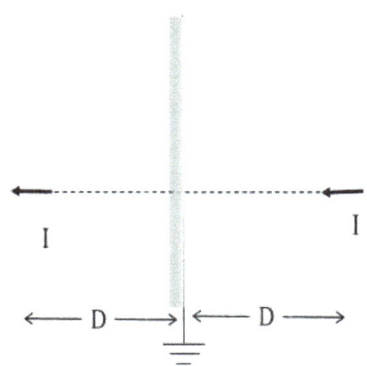

Fig.(3.4) – A current directly perpendicular to the ground plane.

Now, if the current is perpendicular to the plane as shown in Fig.(3.4), it would be equivalent to a positive charge moving toward the ground plane. The negative image charge would of course move toward the ground plane to the right. A negative charge moving to the right means a positive current is moving to the left, as shown.

This concept has been utilized in automobile antenna. Using the car top as ground plane, the antenna is effectively shortened by half.

Fig.(3.5) – A car top antenna, utilizing image current.

Exercise:

3.1 The triangular loop of wire shown in Figure carries a current I = 5 A in the direction shown. The loop is in a uniform magnetic field that has magnitude B = 3 T along the y-axis as shown.
a. Find the force exerted by the magnetic field on each side of the triangle (magnitude and direction).
b. What is the net force on the loop?
c. The loop is pivoted about an axis along ab in the diagram. Calculate the net torque of the loop (magnitude and direction) using $\vec{r} \times \vec{F}$.
d. What is the magnetic moment of the loop?
e. Calculate the torque about the axis ab by using $\vec{\mu} \times \vec{B}$.
f. Describe the direction of the rotation.

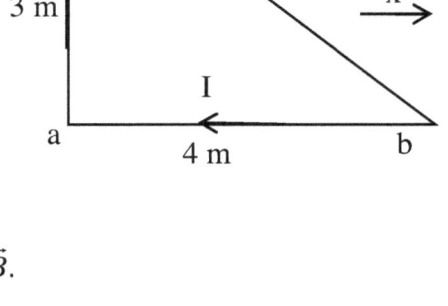

3.2 A <u>long</u> straight copper wire of gauge (AWG) # 15 (diameter = 1.45 mm) is carrying a uniform current density given by J = 8000 (A/m^2). The resistivity of copper is 1.72 x 10^{-6} Ω-cm.
a. What is the magnetic field (vector) created by this current at 0.5 m away from the center of the wire?
b. If an electron is moving in the direction along with the current (as shown in figure) at a speed of 0.01c, would it experience a magnetic force? If so, what is the direction and magnitude of it? If not, why not? [c is speed of light = 3 x 10^8 m/s.]
c. Would the electron exert a magnetic force onto the wire? If so, what's the direction and magnitude of the force? If not, explain why not.

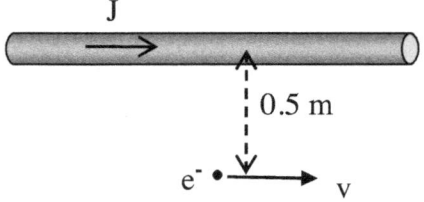

3.3 Point P is at a distance R away from a straight wire of length L carrying a current I as shown in figure. (a) What is the magnetic field vector at point P? (b) If the wire is then extended to a semi-circle and connect to another straight wire of length L (shown in the lower figure), what would the magnetic field at point P become?

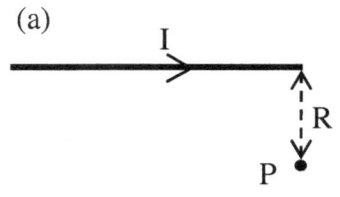
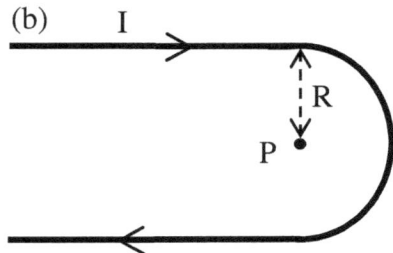

3.4 A semi-circle of radius 2 cm with its center at the origin as shown is carrying a steady current of 5 A. Calculate the magnitude AND direction of the magnetic field at (0,0,4) cm in this coordinate.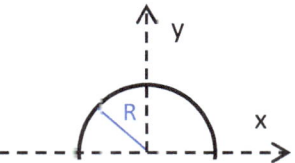

3.5 Derive the magnetic field for an ideal solenoid of length L and has N turns of wire, carrying a current I.

4. Electrodynamics

Chapters 2 and 3 cover the fundamentals of the electrostatic and magnetostatic fields. They can be summarized by 4 static equations.

4.1 Static Fields

$$\begin{cases} \oint \vec{E} \cdot d\vec{a} = \dfrac{Q_t}{\varepsilon_o} \\ \oint \vec{E} \cdot d\vec{\ell} = 0 \\ \oint \vec{B} \cdot d\vec{a} = 0 \\ \oint \vec{B} \cdot d\vec{\ell} = \mu_o I_t \end{cases}$$

Gauss's Law

Conservation

No magnetic charge

Ampere's Law

$$\begin{cases} \nabla \cdot \vec{E} = \dfrac{\rho_t}{\varepsilon_o} \\ \nabla \times \vec{E} = 0 \\ \nabla \cdot \vec{B} = 0 \\ \nabla \times \vec{B} = \mu_o \vec{J}_t \end{cases}$$

Eqn.(4.1)

Eqn.(4.2)

Eqn.(4.3)

Eqn.(4.4)

Eqn.(4.2) is the conservative idea discussed in Section 2.10. It is also equivalent to the Kirchhoff's Loop Law in which the voltage sum around a closed loop is equal to zero.

The third equation is analogous to the Gauss's Law which stated that the net flux is proportional to the charges enclosed. Since there is no magnetic charge, the net magnetic flux through the surface of any enclosed volume is zero, which is the third equation.

The equations on the right are the differential equivalence of the same equations in integral form on the left. For example, Green's theorem relates the area integral to the volume integral for any vector E:

$$\oint \vec{E} \cdot d\vec{a} = \int (\nabla \cdot \vec{E}) dV \qquad \text{Eqn.(4.5)}$$

and the integration volume is enclosed by the surface area on the left. If E is the electric field vector, and together with Gauss's Law:

$$\oint \vec{E} \cdot d\vec{a} = \frac{Q_{total}}{\varepsilon_o}$$

$$\int (\nabla \cdot \vec{E}) dV = \frac{1}{\varepsilon_o} \int \rho \, dV$$

$$\nabla \cdot \vec{E} = \frac{\rho}{\varepsilon_o}$$

which is the same as the equation on the right. The subscript "t" is sometimes added in Q or ρ to remind ourselves that it is the total charge or charge density denoted in these equations. Similarly, Stokes' theorem relates the line integral to the surface integral, for any vector B:

$$\oint \vec{B} \cdot d\vec{\ell} = \int (\nabla \times \vec{B}) \cdot d\vec{a} \qquad \text{Eqn.(4.6)}$$

The area in the integral is bound by the closed loop on the left. Apply this to a magnetic field B, together with the Ampere's Law:

$$\oint \vec{B} \cdot d\vec{\ell} = \mu_o I_t$$

$$\int (\nabla \times \vec{B}) \cdot d\vec{a} = \mu_o \int \vec{J} \cdot d\vec{a}$$

$$\nabla \times \vec{B} = \mu_o \vec{J}$$

where J is the current density defined as the total current through a area divided by the cross-sectional area. Again, subscript "t" is sometimes omitted, but it is still the total current and current density denoted in the equations.

4.2 Faraday's Law

In 1820, Orsted discovered moving charges (current) induced magnetism, which leads to the famous Ampere's Law. A decade later, Michael Faraday in England, and Joseph Henry in USA, independently proved that the reverse is also true. While magnetism itself does not create electricity, changing it does. A more precise definition is first published by Faraday:

$$\Phi = \int \vec{B} \cdot d\vec{a} \qquad \text{Eqn.(4.7)}$$

$$V_{emf} = -\frac{d\Phi}{dt} \qquad \text{Eqn.(4.8)}$$

Equation (4.7) defines the magnetic flux Φ. In mathematics, a vector flux is defined as the dot product of the vector onto the area. Equation (4.8) is the Faraday's Law. The left hand side of the equation historical denoted as ε which stands for the electromotive force (emf). In today's language, it is really the induced voltage that drives the current. V_{emf} is used in this book to signify this is an induced voltage that used to be called emf. The negative in front of the derivative is the essence of Faraday's Law. It stated that the induced emf is to oppose the change of the magnetic flux. This will become more apparent when we discuss Lenz's Law later.

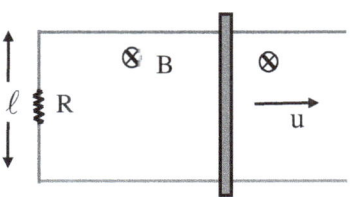

Example (4.1) – A U-shaped wire is submerged in a uniform magnetic field as shown. A conducting bar is placed on top of it and making good electrical contact, and sliding to the right at a constant speed u. Total resistance of the circuit is R. Find the induced current in the loop.

The magnetic flux enclosed by the loop is : $\Phi = \int_0^x B\ell dx = B\ell x$

The induced emf is: $V_{emf} = -\dfrac{d\Phi}{dt} = -B\ell\dfrac{dx}{dt} = -B\ell u$

The magnitude of the induced current: $I = \left|\dfrac{V_{emf}}{R}\right| = \dfrac{B\ell u}{R}$

Faraday's Law is not particularly useful to identify the current direction in a circuit environment. Instead, we should rely on the Lenz's Law.

4.3   Lenz's Law

Lenz's Law dictates the negative sign of the Faraday's Law of induction. It basically stated that the system would induce an emf to oppose the change of the magnetic flux through the system. In other words, the system (or circuit) would induce a current in a way that the magnetic flux would remain constant, at least momentarily.

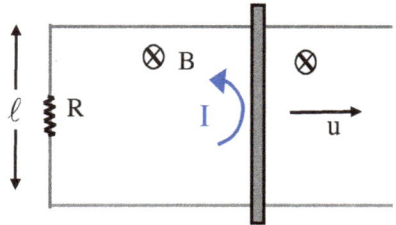

Fig.(4.1) – Current induced as the bar slides on the rail.

In the last example, the magnetic field is pointing into the page. As the bar slides to the right, the area of the loop increases. i.e. the flux increases. Lenz's Law insists to keep the flux unchanged. To do so, a magnetic field is induced to come out of the page, to cancel the increase of the flux. That means the induced current is counter-clockwise as shown, and according to the right-hand-rule.

The right-hand-rule here is: use the thumb as the induced magnetic field, and the curving fingers as the induced current going around the loop. Lenz's Law is all about figuring out the direction of the induced field and current. Let's practice this right-hand-rule in the following examples.

Example (4.2) - Magnet in a tube

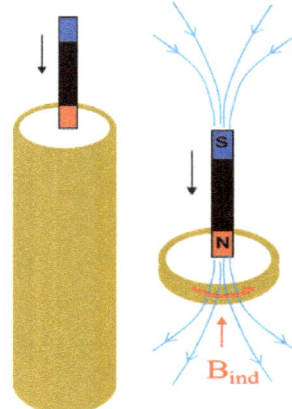

Fig.(4.2) – A magnet thru a tube, and a ring.

There are many online videos on how a magnet slowly "float" down a copper or aluminum tube. It is not because the magnet sticks to the tube (copper and aluminum are not magnetic.) The phenomenon is best explained by Lenz's Law and the induced current (sometimes also called eddy current.)

Let's look at what happens when dropping a bar magnet through a copper ring, as depicted in the right diagram. North pole of the magnet shown is pointing down, so the magnetic field is originated from the lower pole and terminated back in the upper pole.

As the magnet is lowering, the magnetic field intensity increases downward. Lenz's Law said the system (the ring here) would induce a current to maintain a constant magnetic flux (at that moment). Therefore the induced magnetic field has to be pointing up (as shown in the diagram), and the induced current must be going in the direction as shown following the right-hand-rule. Again, with the right thumb pointing straight up, and the curving fingers represent the direction of the circulating induced current. In other words, the ring behaves like a small magnet with the north pole pointing upward. The induced magnetic field push the dropping bar magnet upward so it slows down.

Imagine a tube is made of a series of small rings, the combined effort of the rings pushes the bar magnet upward and significantly slow down the "free-falling" bar magnet. Readers should search for the video and see how amazing this simple concept in work.

Example (4.3) - Magnetic gun?

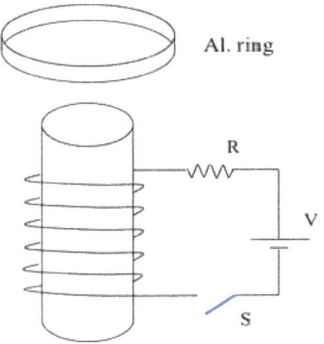

Fig.(4.3a) – Aluminum ring above an electromagnet.

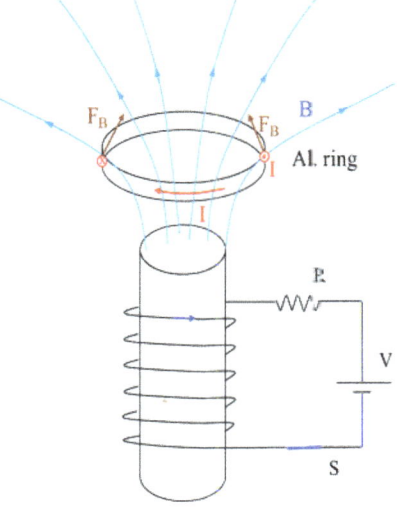

Fig.(4.3b) – After closing the switch.

Place an aluminum ring above an electromagnet as shown in Fig.(4.3a). Close the switch, and the ring will shoot up high. What is happening?

When the switch is closed, a current runs in the electromagnet in the direction as shown in Fig.(4.3b). With the right-hand-rule that curving fingers are the current in the coil, the thumb gives the direction of the B-field which is pointing straight up.

The aluminum ring, started with no magnet flux through it to a sudden increase of flux, would try to oppose this change. It will reduce a magnetic field going down, which requires a reduced current in the direction as shown. i.e., the current comes out of the page on the right side, and loop back into the page on the left side of the ring.

On the right side of the ring, the current is coming out of the page, the magnetic field is pointing to the right and up, as shown. A magnetic force $\vec{F} = I\vec{L} \times \vec{B}$ is pushing the ring up and to the left. On the other side where the current goes into the page, the B-field is pointing up and to the left. The resulting magnetic force is going up and right as shown. The horizontal components of these forces are going to cancel. And the vertical components add up.

If the resultant upward force is greater than the gravitational pull, the ring would shoot up. Aluminum is chosen for this reason because it is the lightest metal with good conductivity.

Example (4.4) - Find the direction of the current in the resistor R shown in Figure at each the following steps: (a) at the instant the switch is closed, (b) after the switch has been closed for several minutes, (c) when the variable resistance r increases, (d) when the circuit containing R moving to the right, away from the other circuit, and (e) at the instant the switch is opened.

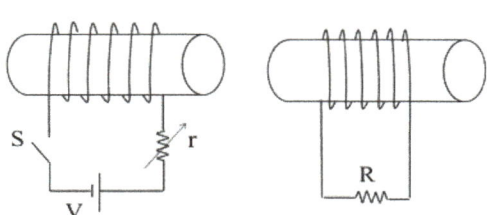

(a) Pay close attention to the winding of the turns. As the switch is closed, the current in the left coil would produce a magnetic field pointing to the right. The 2nd coil on the right would also feel some of this magnetic field. This 2nd coil would then induce a current itself according to Lenz's Law.

The direction of the induced field in the right coil would be pointing to the left to oppose the sudden increase field to the right. With the way the 2nd coil is wound, the induced current in the 2nd circuit would be going from left to right through the resistor R.

(b) This induced current does not last very long. Typically in order of nanoseconds or microseconds at most, depending on the time constant of the circuit; i.e. depends on the inductance and resistance of the circuit. The energy would be dissipated as heat and after a few time constant, there will be practically no induced current unless there are further changes in the magnetic flux. So the answer is NO CURRENT.

(c) As the variable resistance in the 1st circuit increases, its current decreases and so is the magnetic field produced by the 1st coil. This decrease of magnetic field strength is going to trigger an increase in magnetic field to the right in the 2nd coil to make up the difference of the weakening field. To have the induced field pointing to the right, the induced current must be passing the resistor R from right to left.

(d) If the 2nd circuit on the right moves further away from the 1st circuit, the magnetic field (pointing to the right) is going to be weaker, because it is now farther from the source. So the 2nd coil is going to induce a field to the right to make up the difference. Again, it will require a current induced such that the current is going from right to left through the resistor R.

(e) If the switch is now opened and the magnetic field extension produced by the 1st coil into the 2nd coil vanished. The 2nd coil is going to induce a field to make up the different. Again, that would lead up to an induced current passing the resistor R from right to left.

Example (4.5) - A wire, carries a time-varying current I_1, is located at a distance s away from a rectangular loop made of N turns of copper wire of dimensions L and w. The loop is moving away from the line at a velocity v at an angle θ as shown in figure. What is the induced current I_2 in the loop? Use Lenz's Law to confirm the direction of the induced current.

$$\Phi = \int \vec{B} \cdot d\vec{a} = \int_s^{s+w} \frac{\mu_o I_1}{2\pi x} L \, dx = \frac{\mu_o I_o L \sin(\omega t)}{2\pi} \ln\left(\frac{s+w}{s}\right)$$

$$V_{emf} = -N \frac{d\Phi}{dt}$$

$$V_{emf} = -\frac{\omega \mu_o I_o LN \cos(\omega t)}{2\pi} \ln\left(1+\frac{w}{s}\right) - \frac{\mu_o I_o LN \sin(\omega t)}{2\pi} \cdot \frac{-\frac{w}{s^2}\frac{ds}{dt}}{1+\frac{w}{s}}$$

$$I_{induced} = \frac{V_{emf}}{R}$$

$$I_{induced} = -\frac{\omega \mu_o I_o LN \cos(\omega t)}{2\pi R} \ln\left(1+\frac{w}{s}\right) + \frac{\mu_c I_o LN \sin(\omega t) wv \cos\theta}{2\pi R s(s+w)}$$

$I_1 = I_o \sin(\omega t)$

$$R = \rho_{cu}(2L+2w)N / A_{wire\ cross-section}$$

Lenz's Law:

Due to time varying I_1, I_2 is clockwise (+): first term.
Due to the loop moving away, I_2 is counter-clockwise (-): 2nd term.
The correct way to write the final answer is:

$$I_2 = \frac{\mu_o I_o LN}{2\pi R}\left[\omega \cos(\omega t) \ln\left(1+\frac{w}{s}\right) - \frac{wv \cos\theta \sin(\omega t)}{s(s+w)}\right]$$

And yes, it depends on how the I_2 is defined in the diagram.

4.4 Moving conductor

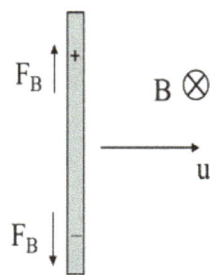

Fig.(4.5) – Moving conductor separate charges.

Faraday's Law and Lenz's Law can be understood from the fundamental magnetic force described in Chapter 3.

Consider a conducting bar, moving in the presence of a magnetic field B, with a constant speed of u, as shown. The positive charge in the conductor would experience a magnetic force $q\vec{v} \times \vec{B}$ which would be pushing the charge upward. Likewise, the force on a negative charge would be downward. In other words, by moving the bar to the right, charges are separated. Positive charges are cumulated on the top, and negatives at the bottom.

When would this charge separation stop? When would the charges be saturated?

As the charge separation occurs, an internal electric field is building up. Recall that electric field is initiated from a positive charge and terminates into a negative charge. So the internal electric field is pointing downward. Equivalently, the relative voltage is positive on top and negative at the bottom.

A positive charge in the conductor now would experience both forces: the magnetic force pushing it up, and the electric force pulling it downward. When these 2 forces are equal and opposite, no more charge separation occurs.

$$\vec{F}_E + \vec{F}_B = 0$$
$$q\vec{E} + q\vec{u} \times \vec{B} = 0$$
$$\vec{E} = -\vec{u} \times \vec{B}$$
$$\frac{V}{\ell} = -uB$$
$$V = -uB\ell$$

This induced voltage is the source of the emf discussed in Faraday's Law. If we allow this moving bar in contact with the open loop circuit in Example (4.1), we can calculate the induced current in the circuit.

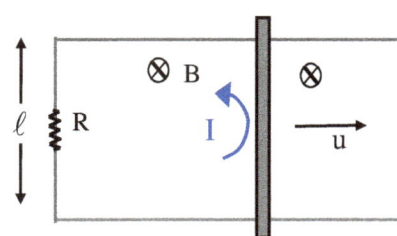

$$I = \frac{|V|}{R} = \frac{uB\ell}{R}$$

which is the same answer obtained in Example (4.1).

The direction is also explained by this voltage, being positive on top and negative a the bottom. The current in the circuit is clearly counter-clockwise as shown.

4.5 Power generator

Power generator and electric motor are very similar. They both involve a permanent magnet, and windings of coils. In case of electric motor discussed in Section 3.8, electrical energy is provided and mechanical torque and work is obtained in return. For power generator, mechanical work is being provided, and electricity is produced as a result. This is one of the major advances in technology using Faraday's Law.

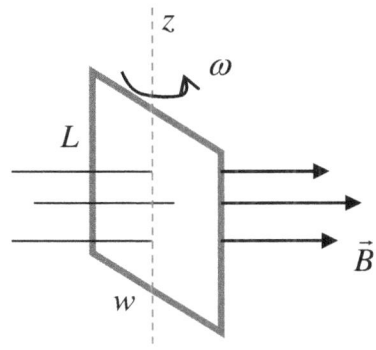

Fig.(4.6) – A rectangular loop rotating in a B-field.

Consider a loop of wire rotating about the z-axis as shown. Faraday's Law simply gives:

$$\Phi = \int \vec{B} \cdot d\vec{a} = BA\cos\theta = BA\cos\omega t$$

$$V_{emf} = -\frac{d\Phi}{dt} = BA\sin\omega t$$

assuming $\theta = 0$ when $t = 0$.

The induced emf is sinosoidual in nature because of the rotational motion. It is therefore more natural to generate AC power than DC historically.

Faraday's Law was a major breakthrough. It stated that Kirchhoff voltage rule is no longer valid if the magnetic flux enclosed by the circuit varies.

$$\Sigma V_i = \oint \vec{E} \cdot d\vec{\ell} = -\frac{d\Phi}{dt}$$

This is why circuits without proper shielding would pick up high frequency noises. To minimize this, all low frequency electronics use twisted wire pair whenever possible.

4.6 Faraday's Law in differential form

Using the Stokes' theorem on the Faraday's equation:

$$\oint \vec{E} \cdot d\vec{\ell} = -\frac{d\Phi}{dt} \qquad (\text{Faraday})$$

$$\int (\nabla \times \vec{E}) \cdot d\vec{a} = -\frac{d}{dt} \int \vec{B} \cdot d\vec{a} \qquad (\text{Stokes})$$

$$\nabla \times \vec{E} = -\frac{\partial \vec{B}}{\partial t}$$

The last equation assumed only the magnetic field change over time. Or equivalently, it is valid for microscopic calculation, or looking at the electromagnetic field locally.

Faraday's Law of induction has so many important applications that it practically changed our civilization. Remote communications were made possible. Telephony, loudspeaker, electric guitar, magnetic brakes, metal detector, and more recently the development of cell phones, blue tooth, and wireless communications. Together with the contribution of Henry, Lenz, and Maxwell....etc, the world has become a very different place.

5. Maxwell's Equations and Boundary Conditions

5.1 Dynamic equations

$$\begin{cases} \nabla \cdot \vec{E} = \dfrac{\rho_t}{\varepsilon_o} & \text{Gauss's Law} \\ \nabla \times \vec{E} = -\dfrac{\partial \vec{B}}{\partial t} & \text{Faraday's Law} \\ \nabla \cdot \vec{B} = 0 & \text{No magnetic charge} \\ \nabla \times \vec{B} = \mu_o \vec{J}_t & \text{Ampere's Law} \end{cases}$$

Eqn.(5.1)
Eqn.(5.2)
Eqn.(5.3)
Eqn.(5.4)

The dynamic equations are essentially the same as the static equations, except with the introduction of Faraday's Law. However, these 4 equations are not consistent. James Clerk Maxwell modified the equations and recognized as the Maxwell's Equations.

5.2 The inconsistence

Take the divergence on both sides of the Faraday's equation:

$$\nabla \times \vec{E} = -\frac{\partial \vec{B}}{\partial t}$$

$$\nabla \cdot \nabla \times \vec{E} = -\nabla \cdot \frac{\partial \vec{B}}{\partial t} = -\frac{\partial}{\partial t}\left(\nabla \cdot \vec{B}\right)$$

The left hand side of the last equation is zero because divergence of a curl is always zero, as discussed in Section 1.6.4. The right hand side of the equation is also equal to zero because $\nabla \cdot \vec{B} = 0$ is one of the dynamic equations. This is perfectly consistence. However, the Ampere's Equation does not seem to work out as well.

$$\nabla \times \vec{B} = \mu_o \vec{J}_t$$

$$\nabla \cdot \nabla \times \vec{B} = \mu_o \left(\nabla \cdot \vec{J}_t\right)$$

On the left hand side, a divergence of a curl is zero. But why is $\nabla \cdot \vec{J}_t = 0$ on the right?

5.3 Continuity Equation

In calculus, if ρ is a function of position and time, and the position $x(t)$ is a dependent variable, then the total derivative of ρ is written as:

$$\frac{d}{dt}\rho(x(t),t) = \frac{d\rho}{dt} = \frac{\partial \rho}{\partial t} + \frac{\partial \rho}{\partial x} \cdot \frac{dx}{dt} = \frac{\partial \rho}{\partial t} + \frac{\partial \rho}{\partial x} \cdot v = \frac{\partial \rho}{\partial t} + \frac{\partial}{\partial x} \cdot (\rho v) \qquad \text{Eqn.(5.5)}$$

The velocity v in Eqn.(5.5) is assumed to be location independent. In 3-dimensions, the velocity is expressed in vector form, and so is its derivative. The total derivative of ρ now reads:

$$\frac{d}{dt}\rho(\vec{r}(t),t) = \frac{\partial \rho}{\partial t} + \nabla \cdot (\rho \vec{v}) \equiv \frac{\partial \rho}{\partial t} + \nabla \cdot \vec{J} \qquad \text{Eqn.(5.6)}$$

where $\vec{J} \equiv \rho \vec{v}$ is the flux of ρ. In fluid dynamics, ρ is the mass density, so J is the mass flux (the amount of mass pass through per unit cross-sectional area per unit time). In thermodynamics, if ρ is the heat energy density, then J is the energy flux. In electrodynamics, ρ would be the charge density, and J is the electrical charge flux (or simply the current density).

In most physical systems, there are conservation laws: conservation of masses, conservation of charges...etc. This translates into a zero total derivative dρ/dt = 0.

$$0 = \frac{\partial \rho}{\partial t} + \nabla \cdot \vec{J} \qquad \text{Eqn.(5.7)}$$

This is referred to as the Continuity Equation. In electrodynamics, it simply says if the charges are moving out of a volume (in form of a current), the charge density inside the volume must be decreasing. It is just another way to write the conservation of charges.

5.4   Maxwell's contribution

Starting with Gauss's Law:

$$\nabla \cdot \vec{E} = \frac{\rho_t}{\varepsilon_o}$$

$$\frac{\partial \rho_t}{\partial t} = \frac{\partial}{\partial t}\left(\varepsilon_o \nabla \cdot \vec{E}\right) = \nabla \cdot \left(\varepsilon_o \frac{\partial \vec{E}}{\partial t}\right)$$

And insert this into the continuity equation:

$$0 = \frac{\partial \rho}{\partial t} + \nabla \cdot \vec{J}_t = \nabla \cdot \left(\vec{J}_t + \varepsilon_o \frac{\partial \vec{E}}{\partial t}\right) \qquad \text{Eqn.(5.8)}$$

So $\nabla \cdot \vec{J}_t \neq 0$, but the divergence of the modified current density in Eqn.(5.8) is.

Faraday added $-\frac{\partial \vec{B}}{\partial t}$ to the $\nabla \times \vec{E}$ equation to take care of the time-varying magnetic field. Likewise, Maxwell modified the Ampere's Law equation with a similar term:

$$\nabla \times \vec{B} = \mu_o \vec{J}_t$$

$$\Rightarrow \nabla \times \vec{B} = \mu_o \left(\vec{J}_t + \varepsilon_o \frac{\partial \vec{E}}{\partial t}\right) \qquad \text{Eqn.(5.9)}$$

Now, take the divergence on this modified Ampere's Law equation:

$$\nabla \cdot \nabla \times \vec{B} = \mu_o \nabla \cdot \left(\vec{J}_t + \varepsilon_o \frac{\partial \vec{E}}{\partial t} \right) = 0$$

The left hand side is a divergence of a curl which is equal to zero. The right hand side is also zero from the continuity equation.

With this modification, the 4 equations (grouped as the Maxwell's Equations) are consistent and read as:

$$\left\{ \begin{array}{l} \nabla \cdot \vec{E} = \dfrac{\rho_t}{\varepsilon_o} \\ \nabla \times \vec{E} = -\dfrac{\partial \vec{B}}{\partial t} \\ \nabla \cdot \vec{B} = 0 \\ \nabla \times \vec{B} = \mu_o \left(\vec{J}_t + \varepsilon_o \dfrac{\partial \vec{E}}{\partial t} \right) \end{array} \right.$$

Gauss's Law — Eqn.(5.10)

Faraday's Law — Eqn.(5.11)

No magnetic charge — Eqn.(5.12)

Ampere / Maxwell — Eqn.(5.13)

Together with the Lorentz Force equation, which is just combining the electric and the magnetic forces: $\vec{F} = q(\vec{E} + \vec{v} \times \vec{B})$, the Maxwell's equations describe all the phenomena in classical electrodynamics, including optics.

5.5 B and H fields

Calculating electromagnetic field in matter can be complicated due to the extra contribution from the molecules in terms of polarization and magnetization. To simplify the task, electric displacement vector D was introduced to only keep track of the free charges, and magnetic field intensity H to account for the free current.

Staightly speaking, B field is called the magnetic induction, or magnetic flux density, while H field is referred to as the magnetic field intensity. Most people, including this book, loosely called them by the same name as "magnetic field".

In electricity, a relative dielectric constant was used to approximation the effect of the polarization induced in the material. In magnetism, permeability is the parameter to describe the magnetic behavior of the material. Away from the atomic scale, these are good approximations to visualize the mean field behavior in the macroscopic world.

$$\vec{B} = \mu \vec{H} = \mu_o \mu_r \vec{H} \qquad \text{Eqn.(5.14)}$$

μ_r is the relative permeability, equals to one for non-magnetic materials.

Similar to ε in dielectric, μ is a tensor (or matrix) to be more exact. What that means is the H field and B field do not have to be in the same direction, especially viewing in microscopic scales.

5.6 Maxwell's Equation of free charges and free current

In terms of free charges and free current, Maxwell's equations can be rewritten as:

$$\left\{\begin{array}{l} \nabla \cdot \vec{D} = \rho_f \\ \nabla \times \vec{E} = -\dfrac{\partial \vec{B}}{\partial t} \\ \nabla \cdot \vec{B} = 0 \\ \nabla \times \vec{H} = \vec{J}_f + \dfrac{\partial \vec{D}}{\partial t} \end{array}\right\} \Leftrightarrow \left\{\begin{array}{l} \oint \vec{D} \cdot d\vec{a} = Q_f \\ \oint \vec{E} \cdot d\vec{\ell} = -\dfrac{\partial}{\partial t} \int \vec{B} \cdot d\vec{a} \\ \oint \vec{B} \cdot d\vec{a} = 0 \\ \oint \vec{H} \cdot d\vec{\ell} = I_f + \dfrac{\partial}{\partial t} \int \vec{D} \cdot d\vec{a} \end{array}\right\} \qquad \text{Eqn.(5.15)}$$

Free charges refers to the net charges that are deposited or induced in the material, as oppose to the bound charges or polarization charges due to the polarization of the molecules. Free current is the real current imposed onto a conductor, not counting the equivalent current due to the change of polarization or magnetization.

5.7 E and D, B and H fields

How to determine whether E or D should be used? And B or H field?
- H relates to free current, which can be measured using an ammeter.
- B highly depends on the material and its history (hysteresis).
- H is easily measureable (via free current) when we build an electromagnet.
- D is related to free charges.
- E is related to the voltage $V = -\int \vec{E} \cdot d\vec{\ell}$, which can be measured with a voltmeter.

Since E and H fields tie to measurements more closely, they are being used more often. But all 4 field descriptions are being used in different disciplines and industries.

5.8 Phasor equations

The Maxwell's Equations in terms of E and H are written as:

$$\left\{\begin{array}{l} \varepsilon \nabla \cdot \vec{E} = \rho_f \\ \nabla \times \vec{E} = -\mu \dfrac{\partial \vec{H}}{\partial t} \\ \nabla \cdot \vec{H} = 0 \\ \nabla \times \vec{H} = \vec{J}_f + \varepsilon \dfrac{\partial \vec{E}}{\partial t} \end{array}\right\} \qquad \text{Eqn.(5.16)}$$

This will be the starting point of the later chapters in electromagnetic waves.

Time-varying electromagnetic fields are functions of space and time. Without getting into the details of separation of variables, one can imagine writing the field equations in Fourier components. In other words, any periodic time function can be written as a sum of harmonic sequence. We can certainly choose to look at one frequency component at a time.

$$\vec{E}(\vec{r},t) = \vec{E}_o(\vec{r})e^{j\omega t}$$
$$\vec{B}(\vec{r},t) = \vec{B}_o(\vec{r})e^{j\omega t} \qquad \text{Eqn.(5.17)}$$

The spatial vectors without the time piece is called the phasor of the functions. Put these in the Maxwell's Equations (5.16), one can pull out a factor of $j\omega$ from the time derivative. And the result is referred to as the phasor equations.

$$\begin{cases} \varepsilon \nabla \cdot \vec{E} = \rho_f \\ \nabla \times \vec{E} = -\mu \dfrac{\partial \vec{H}}{\partial t} \\ \nabla \cdot \vec{H} = 0 \\ \nabla \times \vec{H} = \vec{J}_f + \varepsilon \dfrac{\partial \vec{E}}{\partial t} \end{cases} \Leftrightarrow \begin{cases} \nabla \cdot \vec{E} = \dfrac{\rho_f}{\varepsilon} \\ \nabla \times \vec{E} = -j\omega\mu \vec{H} \\ \nabla \cdot \vec{H} = 0 \\ \nabla \times \vec{H} = \vec{J}_f + j\omega\varepsilon\vec{E} \end{cases} \qquad \text{Eqn.(5.18)}$$

The Maxwell's Equations relate electric field with the magnetic field. The phasor equations make the coupling between E and H fields through simple algebraic manipulations.

Example (5.1) - $\vec{H}(\vec{r},t) = \hat{x}0.01\cos(9\cdot 10^9 t + 30z)$ A/m in vacuum with no electrical current. Find E.

In general, the number in front of t is the angular frequency ω, and in front of the spatial parameter is the wave number in that direction. (More details in the plane wave chapter.)

$$\nabla \times \vec{H} = j\omega\varepsilon\vec{E}$$

$$\nabla \times \vec{H} = \begin{vmatrix} \hat{x} & \hat{y} & \hat{z} \\ \dfrac{\partial}{\partial x} & \dfrac{\partial}{\partial y} & \dfrac{\partial}{\partial z} \\ 0.01\cos(\omega t + 30z) & 0 & 0 \end{vmatrix} = -\hat{y}0.01(30)\sin(\omega t + 30z)$$

$$j\omega\varepsilon\vec{E} = -\hat{y}0.3\sin(\omega t + 30z)$$

$$\vec{E} = \dfrac{-\hat{y}0.3\sin(\omega t + 30z)}{j\omega\varepsilon_o} = j\dfrac{\hat{y}0.3\sin(\omega t + 30z)}{(9\cdot 10^9)(10^{-9}/36\pi)} = j\hat{y}(1.2\pi)\sin(\omega t + 30z)$$

The imaginary number j also means 90 degrees ahead in phase. So, $j\sin(\theta)=\cos(\theta)$.

$$\vec{E} = \hat{y}(3.77)\cos(9\cdot 10^9 t + 30z) \quad \text{V/m.}$$

5.9 Boundary conditions

It was shown in Chapter 2.8 that electric field is affected by the presence of a conductor. In particular, electric field is bent toward the conductor so it is always perpendicular to the surface of a conductor. In other materials, the field will bend somewhat when it is passing through, much the same way the light refracts through dielectrics. Let's examine the field behaviors at the boundary.

5.9.1 Normal component of D field

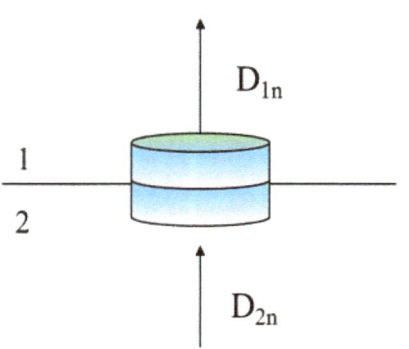

For media 1 and 2 as shown, imagine a very thin pill box cutting through the boundary. The normal components of the D fields are related using Gauss's Law:

$$\oint \vec{D} \cdot d\vec{a} = Q_f$$

$$D_{1n} A - D_{2n} A = \sigma_s A$$

$$D_{1n} - D_{2n} = \sigma_s \quad \text{Eqn.(5.19)}$$

σ_s is the surface charge density.

The pill box is so infinitesimally thin that there will be no volume charge enclosed. If both media are non-conducting dielectric, $D_{1n} = D_{2n}$. If medium 2 is a conductor, for example, then D_2 would be zero because there is no electric field (or D field) inside a good conductor. But there is a surface charge because free charges always stay on the surface of a good conductor (Section 2.8.2), therefore, $D_{1n} = \sigma_s$. E and D fields are related by ε, so one can write the relationship for the normal component of electric fields as well:

$$\varepsilon_o (\varepsilon_{r1} E_{1n} - \varepsilon_{r2} E_{2n}) = \sigma_s \quad \text{Eqn.(5.20)}$$

5.9.2 Tangential component of E field

To look at the tangential components of the fields, we draw a thin contour just above and below the surface of the boundary.

$$\oint \vec{E} \cdot d\vec{\ell} = -\frac{\partial}{\partial t} \int \vec{B} \cdot d\vec{a} = 0$$

$$E_{1t} \ell - E_{2t} \ell = 0$$

$$E_{1t} = E_{2t} \quad \text{Eqn.(5.21)}$$

The contour is so thin that the area enclosed is practically zero. So the area integral would be zero for any finite B field. i.e., the tangential component of E field is always continuous.

5.9.3 Normal component of B field

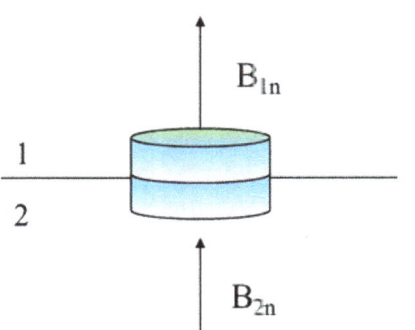

To examine the normal component of the B field, we use the pill box again.

$$\oint \vec{B} \cdot d\vec{a} = 0$$

$$B_{1n} A - B_{2n} A = 0$$

$$B_{1n} = B_{2n} \qquad \text{Eqn.(5.22)}$$

Normal component of B field is always continuous.

5.9.4 Tangential component of H field

$$\oint \vec{H} \cdot d\vec{\ell} = I_f - \frac{\partial}{\partial t} \int \vec{D} \cdot d\vec{a}^{\,0}$$

$$H_{1t} \ell - H_{2t} \ell = J_s \ell$$

$$\hat{n}_2 \times (\vec{H}_{1t} - \vec{H}_{2t}) = \vec{J}_s$$

$$\hat{n}_2 \times (\vec{H}_1 - \vec{H}_2) = \vec{J}_s \qquad \text{Eqn.(5.23)}$$

Again, the contour chosen is so thin that the area is practically zero. Therefore, the area integral is zero. J_s is the surface current density. Free charges stay on the surface of a conductor, so do free currents. H field ties to free current, so one can also assume H is also zero inside a conductor. Again, this is not exactly true microscopically except for superconductors. But for simplicity and discussions in this book, H = 0 inside a good conductor is a good approximation.

Example (5.2) – Image current.

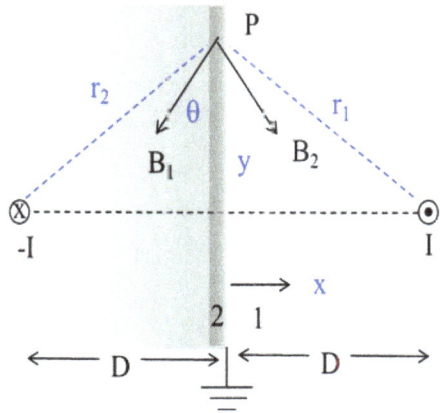

An infinite wire carrying a current I is place in front of a ground plane as shown. Using the image current method in Chapter 3.10, find the surface current density at point P.

From Ampere's Law: -Example (3.5)

$$B_1 = \frac{\mu_o I}{2\pi r_1} = B_2 = \frac{\mu_o I}{2\pi r_2}$$

At point P, the normal components cancel, and the tangential components add up:

$$B_{y_{total}} = -2B_1 \cos\theta = -\frac{\mu_o I y}{\pi r^2} = -\frac{\mu_o I y}{\pi(y^2 + D^2)}$$

Note: here at point P, $B_{1n} = B_{2n} = 0$. And

$$\vec{J}_s = \hat{n}_2 \times (\vec{H}_{1t} - \vec{H}_{2t}) = \hat{x} \times \left(\frac{\vec{B}_{1t}}{\mu_o} - 0\right) = \hat{x} \times \left(-\hat{y}\frac{\mu_o I y}{\pi(y^2 + D^2)}\right) = -\hat{z}\frac{I y}{\pi(y^2 + D^2)} \otimes$$

The induced surface current on the ground plane is in the –z direction, which is into the page in the figure shown.

Example (5.3) - A dielectric boundary defined by the xy-plane (i.e., z = 0), separates the air (z < 0) from a dielectric (z > 0) with a relative dielectric constant of 4. If the electric field in air right by the boundary (z = 0⁻) is given by: $\vec{E}_1 = \hat{x} - 2\hat{y} + 4\hat{z}$, find E₂ in the dielectric (z = 0⁺).

Both media are dielectric with no surface charge:
$$D_{1n} - D_{2n} = 0$$
$$E_{1n} = \varepsilon_{r2} E_{2n}$$
$$E_{2z} = E_{2n} = \frac{E_{1n}}{\varepsilon_{r2}} = \frac{4}{4} = 1$$

Tangential components continuous: $E_{2x} = E_{1x} = 1$, $E_{2y} = E_{1y} = -2$.

Altogether, $\vec{E}_2 = \hat{x} - 2\hat{y} + \hat{z}$.

6 Plane electromagnetic waves

Maxwell's equations (including the Faraday's Law) clearly indicated that time-varying electric field creates magnetism, and time-varying magnetic field produces electricity. Together with Einstein's special relativity, one can conclude that electricity and magnetism cannot exist by itself without the other. What appears to be purely electrostatic to an observer could appear to be purely magnetostatic in another inertia reference frame. In fact, Einstein's first publication on the subject published in 1905 was translated to "On the electrodynamics of moving bodies".

One of the interesting consequences of Maxwell's equations is that a time-varying field will produce an electromagnetic wave. In other words, an AC circuit would radiate electromagnetic waves around it. Most electronic products, especially at higher frequency, are required to shield off its radiation to avoid electromagnetic spectrum pollution. In this chapter, we explain how this phenomenon comes about.

6.1 Source-free Maxwell's Equations

Let us begin with the simplest scenario where there is no free charge or current in the medium, and investigate how the time-varying electromagnetic field behaves. With no charge or current source, the Maxwell's equations of E and H fields from Eqn.(5.16) becomes"

$$\begin{cases} \varepsilon \nabla \cdot \vec{E} = \rho_f \\ \nabla \times \vec{E} = -\mu \frac{\partial \vec{H}}{\partial t} \\ \nabla \cdot \vec{H} = 0 \\ \nabla \times \vec{H} = \vec{J}_f + \varepsilon \frac{\partial \vec{E}}{\partial t} \end{cases} \begin{matrix} \rho_f = 0 \\ \Rightarrow \Rightarrow \Rightarrow \Rightarrow \Rightarrow \\ \vec{J}_f = 0 \end{matrix} \begin{cases} \nabla \cdot \vec{E} = 0 \\ \nabla \times \vec{E} = -\mu \frac{\partial \vec{H}}{\partial t} \\ \nabla \cdot \vec{H} = 0 \\ \nabla \times \vec{H} = \varepsilon \frac{\partial \vec{E}}{\partial t} \end{cases} \qquad \text{Eqn.(6.1)}$$

The two non-zero equations couple the electric and magnetic fields. In particular, these equations can be combined to produce a wave equation. For example,

$$\nabla \times \vec{E} = -\mu \frac{\partial \vec{H}}{\partial t}$$

$$\nabla \times (\nabla \times \vec{E}) = \nabla \times \left(-\mu \frac{\partial \vec{H}}{\partial t}\right) = -\mu \frac{\partial}{\partial t}(\nabla \times \vec{H}) = -\mu \frac{\partial}{\partial t}\left(\varepsilon \frac{\partial \vec{E}}{\partial t}\right) = -\mu\varepsilon \frac{\partial^2 \vec{E}}{\partial t^2} \qquad \text{Eqn.(6.2)}$$

But from Example 1.7, for any vector E: $\nabla \times (\nabla \times \vec{E}) = \nabla(\nabla \cdot \vec{E}) - \nabla^2 \vec{E}$
Together with the first equation that $\nabla \cdot \vec{E} = 0$, we have $\nabla \times (\nabla \times \vec{E}) = -\nabla^2 \vec{E}$

Therefore, $\nabla^2 \vec{E} = \mu\varepsilon \frac{\partial^2 \vec{E}}{\partial t^2}$ $\qquad$ Eqn.(6.3)

This is a wave equation, and the velocity of the wave is $\frac{1}{\sqrt{\mu\varepsilon}}$. Properties of traveling wave equation are reviewed in Appendix D.

Similarly, if we start with $\nabla \times (\nabla \times \vec{H})$, we would end up with $\nabla^2 \vec{H} = \mu\varepsilon \frac{\partial^2 \vec{H}}{\partial t^2}$.

Of course, the electric field and magnetic field are coupled together. One wave cannot exist without the other. In other words, if the electromagnetic field is time-varying, such as from a AC source in a circuit, there will be electromagnetic wave come out of the source or circuit. For this reason, all electronic products (especially in radio frequency) are under strict regulation on electromagnetic radiation, and they are typically enclosed by conducting ground shields.

In vacuum, the wave velocity is $\frac{1}{\sqrt{\mu_o \varepsilon_o}} = 3 \cdot 10^8 \, m/s = c$.

If we are far away from the source of radiation, the wave front could be approximated as a plane perpendicular to the direction of propagation. Unless we are looking at the wave behavior very close to the radiating element, all subsequent discussions in the book will focus on the electromagnetic waves in the plane wave limit.

<u>Example (6.1)</u> - $\vec{E}(\vec{r},t) = \hat{x} E_o e^{j(\omega t - kz)}$ is found in vacuum with no charge or current source. What is the corresponding $\vec{H}(\vec{r},t)$?

We can use the phasor equation (5.18), and set ρ = 0 and J = 0 for source-free environment.

$$\begin{cases} \nabla \cdot \vec{E} = \frac{\rho_f}{\varepsilon} \\ \nabla \times \vec{E} = -j\omega\mu\vec{H} \\ \nabla \cdot \vec{H} = 0 \\ \nabla \times \vec{H} = \vec{J}_f + j\omega\varepsilon\vec{E} \end{cases} \Leftrightarrow \begin{cases} \nabla \cdot \vec{E} = 0 \\ \nabla \times \vec{E} = -j\omega\mu\vec{H} \\ \nabla \cdot \vec{H} = 0 \\ \nabla \times \vec{H} = j\omega\varepsilon\vec{E} \end{cases} \qquad \text{Eqn.(6.4)}$$

Clearly from here, E and H fields are directly coupled. $\nabla \times \vec{E} = -j\omega\mu\vec{H}$

$$\nabla \times \vec{E} = \begin{vmatrix} \hat{x} & \hat{y} & \hat{z} \\ \frac{\partial}{\partial x} & \frac{\partial}{\partial y} & \frac{\partial}{\partial z} \\ E_o e^{j(\omega t - kz)} & 0 & 0 \end{vmatrix} = \hat{y}\{-jk E_o e^{j(\omega t - kz)}\} = -j\omega\mu\vec{H}$$

$$\vec{H}(\vec{r},t) = \hat{y} \frac{k E_o}{\omega\mu} e^{j(\omega t - kz)}$$

$\frac{k}{\omega\mu} = \frac{1}{f\lambda\mu} = \frac{1}{v\mu} = \frac{\sqrt{\mu\varepsilon}}{\mu} = \sqrt{\frac{\varepsilon}{\mu}} \equiv \frac{1}{\eta}$ η defined here is called the wave impedance.

$$\vec{H}(\vec{r},t) = \hat{y} \frac{E_o}{\eta} e^{j(\omega t - kz)}$$

$$\vec{B}(\vec{r},t) = \mu\vec{H} = \hat{y} \frac{\mu E_o}{v\mu} e^{j(\omega t - kz)} = \hat{y} \frac{E_o}{v} e^{j(\omega t - kz)}$$

6.2 Properties of plane electromagnetic waves

Example (6.1) highlighted some interesting properties of electromagnetic waves. It started off with a very general plane wave, propagating in the +z direction, ended with a H field propagating in the same direction. Let us examine the pair of equations again and identify some general features.

$$\vec{E}(\vec{r},t) = \hat{x} E_o e^{j(\omega t - kz)}$$

$$\vec{H}(\vec{r},t) = \hat{y} \frac{E_o}{\eta} e^{j(\omega t - kz)}$$

6.2.1 E and H are always in phase.

This is true as long as the medium the wave is propagating in is lossless or with very low dielectric loss. Basically, if E is given, H would have the same phase information.

6.2.2 $\vec{E} \perp \vec{H} \perp \vec{k}$ and they follow the right-hand-rule and are in cyclic permutation.

In this example, E is in the +x direction, H is in the +y direction, and k is along the +z direction. In fact, one can write, in general: $\hat{k} = \hat{E} \times \hat{H}$

6.2.3 The ratio of E_o / H_o is the wave impedance $\eta \equiv \sqrt{\frac{\mu}{\varepsilon}}$.

In vacuum, $\eta_o = \sqrt{\frac{\mu_o}{\varepsilon_o}} = \sqrt{\frac{4\pi \cdot 10^{-7}}{\frac{1}{36\pi} \cdot 10^{-9}}} = 120\pi = 377\Omega$

With these 3 properties, one can easily write the H-field with a given E-field, and vice versa.

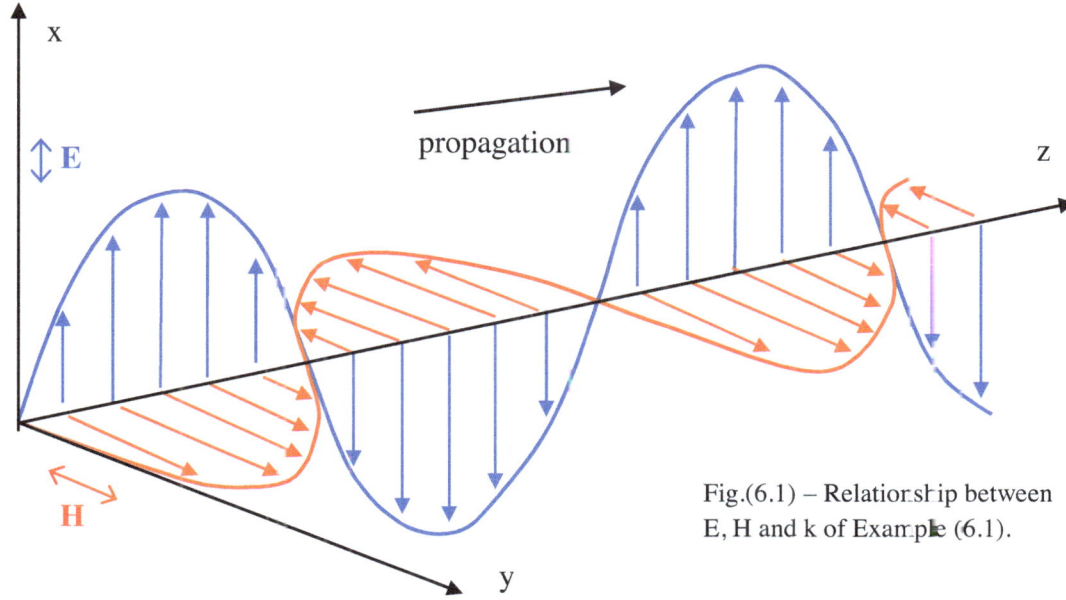

Fig.(6.1) – Relationship between E, H and k of Example (6.1).

6.2.4 The ratio of E_o / B_o is the phase velocity $v = \sqrt{\dfrac{1}{\mu \varepsilon}}$.

In vacuum, the phase velocity is just the speed of light = c. In general, one can write:

$$v = \sqrt{\frac{1}{\mu\varepsilon}} = \sqrt{\frac{1}{\mu_o \mu_r \varepsilon_o \varepsilon_r}} = \frac{c}{\sqrt{\mu_r \varepsilon_r}}.$$

For non-magnetic material, the relative permeability $\mu_r = 1$, and $v = \dfrac{c}{\sqrt{\varepsilon_r}} = \dfrac{c}{n}$.

Here, n is the index of refraction, not to be confused with the wave impedance η.

<u>Example (6.2)</u> - Given $\vec{H}(\vec{r},t) = \hat{z} 0.02 \cos(10^9 t + 40y)$ A/m, in a non-magnetic medium. Find the frequency, wavelength, phase velocity of the wave, the dielectric constant of the medium, and the corresponding electric field.

$$f = \frac{\omega}{2\pi} = \frac{10^9}{2\pi} = 159.2 MHz$$

$$\lambda = \frac{2\pi}{k} = \frac{2\pi}{40} = 0.157 m$$

$$v = f\lambda = \frac{\omega}{k} = \frac{10^9}{40} = 2.5 \cdot 10^7 m/s$$

$$\frac{c}{v} = \frac{3 \cdot 10^8}{2.5 \cdot 10^7} = 12 = n = \sqrt{\varepsilon_r}$$

$$\varepsilon_r = 144$$

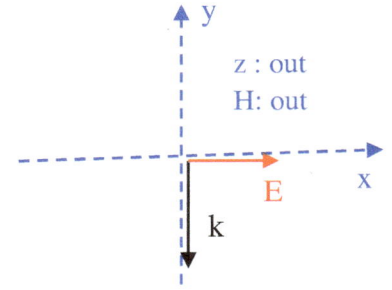

$$\eta = \sqrt{\frac{\mu}{\varepsilon}} = \sqrt{\frac{\mu_o}{\varepsilon_o \varepsilon_r}} = \frac{\eta_o}{\sqrt{\varepsilon_r}} = \frac{377}{12} = 31.4$$

$$\frac{E_o}{H_o} = \frac{E_o}{0.02} = 31.4$$

$$E_o = 94.25(0.02) = 0.63$$

$$\vec{E}(\vec{r},t) = \hat{x} 0.63 \cos(10^9 t + 40y) \quad \text{V/m}$$

This is a very general approach and can be applied to a wide range of situations, including those with E, H or k not pointing along an axis. Of course, if the direction cannot be clearly illustrated in a diagram, calculation (from Maxwell's Equations) would be easier.

Example (6.3) - Given that $\vec{H}(\vec{r},t) = (\hat{x}0.01 - \hat{y}0.02)\sin(10^9 t + \beta z - 0.1)$ A/m in vacuum, what is the corresponding E-field?

$$\beta = \frac{\omega}{c} = \frac{10^9}{3 \cdot 10^8} = \frac{10}{3} = 3.33 m^{-1}$$

(a) We can use vector notation to treat it altogether:

In phase: $\vec{E}(\vec{r},t) = \vec{E}_o \sin(10^9 t + \beta z - 0.1)$

Right-hand-rule: $\vec{E} \perp \vec{H} \perp \vec{k}$

$$\hat{k} = \hat{E} \times \hat{H}$$

$$\hat{E} = \frac{2\hat{x} + \hat{y}}{\sqrt{5}} = 0.894\hat{x} + 0.447\hat{y}$$

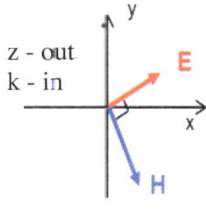

$E_o/H_o = \eta$:

$$H_o = \sqrt{0.01^2 + 0.02^2} = 0.022$$

$$E_o = \eta_o H_o = (377)(0.022) = 8.3 V/m$$

$$\vec{E}(\vec{r},t) = 8.3(\hat{x}0.894 + \hat{y}0.447)\sin(10^9 t + 3.33z - 0.1)$$

$$\vec{E}(\vec{r},t) = (\hat{x}7.4 + \hat{y}3.7)\sin(10^9 t + 3.33z - 0.1)$$

(b) Or, we can separate this into 2 different parts as if they are 2 separate problems:

$$\vec{H}(\vec{r},t) = \hat{x}0.01\sin(10^9 t + \beta z - 0.1) - \hat{y}0.02\sin(10^9 t + \beta z - 0.1)$$

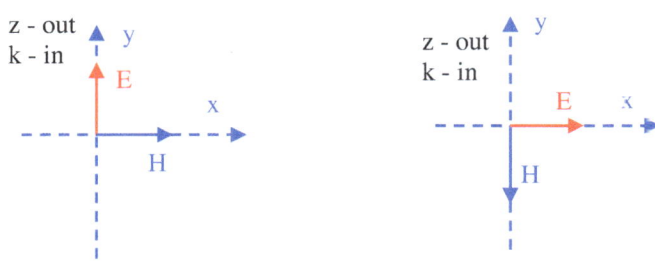

$$\vec{E}(\vec{r},t) = \hat{y}(0.01\eta_o)\sin(10^9 t + \beta z - 0.1) + \hat{x}(0.02\eta_o)\sin(10^9 t + \beta z - 0.1)$$

$$\vec{E}(\vec{r},t) = (0.01\hat{y} + 0.02\hat{x})\eta_o \sin(10^9 t + \beta z - 0.1)$$

$$\vec{E}(\vec{r},t) = (\hat{y} + 2\hat{x})3.77\sin(10^9 t + 3.33z - 0.1) \quad \text{V/m}$$

In this case, method (b) would be easier. But sometimes we cannot avoid using the first vector method, as illustrated in the following example.

Example (6.4) - Given that $\vec{E}(\vec{r},t) = \hat{x} 20\cos(10^8 t + 3z - y)$ V/m in a non-magnetic material, find H-field.

Note: We cannot assume the medium is a vacuum especially if ω and k are given.

In phase: $\vec{H}(\vec{r},t) = \vec{H}_o \cos(10^8 t + 3z - y)$

Right-hand-rule: $\vec{k} = -3\hat{z} + \hat{y}$ (given)

$$\hat{H} = \frac{-3\hat{y} - \hat{z}}{\sqrt{3^2 + 1^2}} = -\frac{3\hat{y} + \hat{z}}{\sqrt{10}}$$

$E_o/H_o = \eta$:

$$k = \sqrt{3^2 + 1^2} = \sqrt{10}$$

$$v = \frac{\omega}{k} = \frac{10^8}{\sqrt{10}} = \frac{c}{n} = \frac{3\cdot 10^8}{n}$$

$$n = 3\sqrt{10}$$

$$\eta = \sqrt{\frac{\mu}{\varepsilon}} = \sqrt{\frac{\mu_r \mu_o}{\varepsilon_r \varepsilon_o}} = \sqrt{\frac{\mu_o}{\varepsilon_o}} \cdot \sqrt{\frac{1}{\varepsilon_r}} = \frac{\eta_o}{n} = \frac{377}{3\sqrt{10}} = 39.7$$

$$H_o = \frac{E_o}{\eta} = \frac{20}{39.7} = 0.5$$

Altogether,

$$\vec{H}(\vec{r},t) = -\frac{(3\hat{y} + \hat{z})}{\sqrt{10}}(0.5)\cos(10^8 t + 3z - y) = -(0.475\hat{y} + 0.158\hat{z})\cos(10^8 t + 3z - y) \quad \text{A/m}$$

6.3 Direct cross-product

In section 6.1, we learned how to relate the E and H field by taking the curl in the Maxwell's Equations as in Eqn.(6.4). In section 6.2, we simply write the E or H field by using the 3 properties of the fields derived from, of course, the Maxwell's Equations. Yet, one can rewrite the 3 properties in a more direct way by stating:

$$\vec{H} = \frac{1}{\eta}\hat{k} \times \vec{E}$$

$$\vec{E} = \eta \vec{H} \times \hat{k} \quad \quad \quad \text{Eqn.(6.5)}$$

This is just a compact way to summarize the 3 properties listed in Section 6.2. So effective, we now have 3 different ways to calculate E from H, or vice versa. Let us revisit Example (6.4) and apply this formula to get the answer instead.

Example (6.5) - Apply Eqn.(6.5) to Example (6.4) and find H-field.

$$\vec{E}(\vec{r},t) = \hat{x}20\cos(10^8 t + 3z - y)$$

$$\vec{k} = -3\hat{z} + \hat{y}$$

$$\hat{k} = \frac{-3\hat{z} + \hat{y}}{\sqrt{3^2 + 1^2}} = \frac{-3\hat{z} + \hat{y}}{\sqrt{10}}$$

$$\eta = \sqrt{\frac{\mu}{\varepsilon}} = \sqrt{\frac{\mu_r \mu_o}{\varepsilon_r \varepsilon_o}} = \sqrt{\frac{\mu_o}{\varepsilon_o}} \cdot \sqrt{\frac{1}{\varepsilon_r}} = \frac{\eta_o}{n} = \frac{377}{3\sqrt{10}}$$

$$\vec{H} = \frac{1}{\eta}\hat{k} \times \vec{E} = \frac{3\sqrt{10}}{377}\left(\frac{-3\hat{z} + \hat{y}}{\sqrt{10}}\right) \times \hat{x}20\cos(10^8 t + 3z - y)$$

$$\vec{H} = \frac{60}{377}(-3\hat{y} - \hat{z})\cos(10^8 t + 3z - y) = -(0.477\hat{y} + 0.159\hat{z})\cos(10^8 t + 3z - y) \quad \text{A/m}$$

6.4 Poynting Theorem

The Poynting Theorem is basically a work-energy theorem for electrodynamics. Consider the work done on a small charge δq by electromagnetic force:

$$dW = \vec{F} \cdot d\vec{\ell} = \delta q(\vec{E} + \vec{v} \times \vec{B}) \cdot \vec{v} dt$$

$$dW = \delta q \vec{E} \cdot \vec{v} dt$$

$$dW = (\rho \delta V)\vec{E} \cdot \vec{v} dt = \delta V \vec{E} \cdot \vec{J} dt$$

$$\frac{dW}{dt} = \int \vec{E} \cdot \vec{J} dV$$

$$\nabla \times \vec{H} = \vec{J}_f + \varepsilon \frac{\partial \vec{E}}{\partial t}$$

$$\vec{E} \cdot \vec{J} = \vec{E} \cdot \nabla \times \vec{H} - \varepsilon \vec{E} \cdot \frac{\partial \vec{E}}{\partial t}$$

$$\nabla \cdot (\vec{E} \times \vec{H}) = \vec{H} \cdot \nabla \times \vec{E} - \vec{E} \cdot \nabla \times \vec{H}$$

$$\vec{E} \cdot \nabla \times \vec{H} = \vec{H} \cdot \left(-\mu \frac{\partial \vec{H}}{\partial t}\right) - \nabla \cdot (\vec{E} \times \vec{H})$$

$$\vec{E} \cdot \vec{J} = -\mu \vec{H} \cdot \frac{\partial \vec{H}}{\partial t} - \nabla \cdot (\vec{E} \times \vec{H}) - \varepsilon \vec{E} \cdot \frac{\partial \vec{E}}{\partial t}$$

$$\vec{E} \cdot \vec{J} = -\frac{\mu}{2}\frac{\partial \vec{H}^2}{\partial t} - \frac{\varepsilon}{2}\frac{\partial \vec{E}^2}{\partial t} - \nabla \cdot (\vec{E} \times \vec{H})$$

$$\frac{dW}{dt} = -\int \frac{1}{2}\left(\mu \frac{\partial \vec{H}^2}{\partial t} + \varepsilon \frac{\partial \vec{E}^2}{\partial t}\right)dV - \int \nabla \cdot (\vec{E} \times \vec{H}) dV$$

Remember magnetic energy density = μH^2 which is the magnetic energy per volume, and electric energy density = εE^2, and use the Divergence Theorem to rewrite the last term:

$$\frac{dW}{dt} = -\int \frac{1}{2}\left(\mu \frac{\partial \vec{H}^2}{\partial t} + \varepsilon_o \frac{\partial \vec{E}^2}{\partial t}\right) dV - \oint (\vec{E} \times \vec{H}) \cdot d\vec{a}$$

$$\frac{dW}{dt} \equiv -\frac{\partial \overline{U}_{EM}}{\partial t} - \oint \vec{S} \cdot d\vec{a}$$

$$-\frac{\partial \overline{U}_{EM}}{\partial t} = \frac{dW}{dt} + \oint \vec{S} \cdot d\vec{a} \qquad \text{Eqn.(6.6)}$$

where $\vec{S} \equiv \vec{E} \times \vec{H}$ is called the Poynting Vector which has the direction of the wave propagation, and has the magnitude equals to the power intensity (= power per unit area).

$\oint \vec{S} \cdot d\vec{a}$ represents the electromagnetic power flow out of the enclosed area defined by the close loop integral. Equation (6.6) basically says that the decrease of the electromagnetic energy is either to do work on the charge or radiate out of the area.

The time-average power intensity can be written as:

$$\vec{S}_{ave} \equiv \frac{1}{2} \Re e[\vec{E} \times \vec{H}^*]$$

$$S_{ave} = \frac{|E_o|^2}{2\eta} = \frac{\eta}{2}|H_o|^2 \qquad \text{Eqn.(6.7)}$$

Remember, η is the wave impedance. This power intensity equation (6.7) is similar to the more familiar power equations of:

$$P_{ave} = \frac{1}{2}\frac{V^2}{Z} = \frac{1}{2} I^2 Z \qquad \text{Eqn.(6.8)}$$

Again, the relation between V and E_o, and between I and H_o are apparent.

6.5 Polarization

The behavior of an electromagnetic wave when it enters from one region to another, or when it reflects from a surface, depends on the orientation of the fields with respect to the surface. To study the effect of the field orientation, people defined the polarization of the wave by the orientation of its electric field vector.

6.5.1 Linear polarization

If the electric field remains oscillating in one-direction, it is said to have a linear polarization. For example,

$\vec{E}_1(\vec{r},t) = \hat{x} E_{o1} \cos(\omega t - \beta z)$ has a field oscillating in +/- x-axis, and is propagating along the z-axis (out of the paper). This wave is a linearly polarized wave, or more specifically, a horizontally polarized wave.

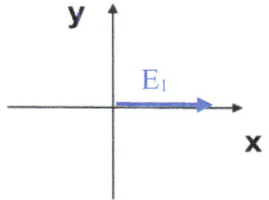

Fig.(6.2) – Horizontal Polarization.

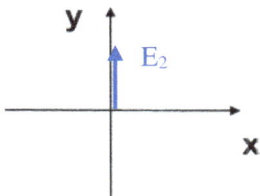

Likewise, $\vec{E}_2(\vec{r},t) = \hat{y} E_{o2} \cos(\omega t - \beta z)$ also propagates along the z-axis, but oscillates in the +/- y-axis, is also a linearly polarized wave. It is vertically polarized.

Fig.(6.3) – Vertical Polarization.

A linear combination of 2 linearly polarized waves is still linearly polarized. $\vec{E}_{total} = \vec{E}_1 + \vec{E}_2$

If E_1 and E_2 have the same amplitude, then E_{total} would be oscillating along the 45-degrees line. But it is still a linearly polarized wave.

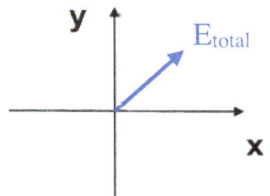

Fig.(6.4) – Linear Polarization results from adding 2 linearly polarized waves.

Just like any 2-dimensional vector can be written as a linear combination of 2 vectors that span the 2-dimensional space (such as the $\hat{x}$ and $\hat{y}$), any electromagnetic wave can be written as a linear combination of a horizontal and a vertical wave.

6.5.2   Circular polarization

If E_1 is a horizontally polarized wave and E_2 is a vertically polarized wave as stated before, but have a 90-degree phase shift with respect to each other, then the combined wave is circularly polarized. For example:

$$\vec{E}_1(\vec{r},t) = \hat{x} E_o \cos(\omega t - \beta z)$$
$$\vec{E}_2(\vec{r},t) = \hat{y} E_o \sin(\omega t - \beta z)$$
$$\vec{E}_{total}(\vec{r},t) = \hat{x} E_o \cos(\omega t - \beta z) + \hat{y} E_o \sin(\omega t - \beta z)$$
$$\vec{E}_{total}(\vec{r},t) = \hat{x} E_o e^{j(\omega t - \beta z)} + \hat{y} E_o e^{j(\omega t - \beta z - \pi/2)}$$
$$\vec{E}_{total}(\vec{r},t) = (\hat{x} - j\hat{y}) E_o e^{j(\omega t - \beta z)}$$

Eqn.(6.9)

The last 3 equations are essentially the same, just different ways to express the same information. To illustrate the circular nature of the polarization, it would be useful to consider the electric field orientation at t = 0 around the origin. In this case, let us examine the equation again: $\vec{E}(\vec{r},t) = \hat{x} E_o \cos(\omega t - \beta z) + \hat{y} E_o \sin(\omega t - \beta z)$

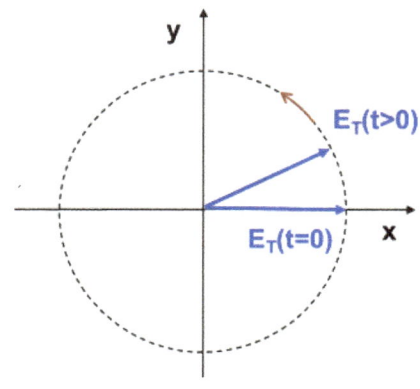

Fig.(6.5) – RHCP

At origin z = 0 and t = 0, the y-component becomes zero. Electric field is pointing in the x-direction as shown in figure. As time increases, the y-component starts picking up amplitude, and at the same time, the x-component amplitude begins to drop. In other words, the electric field at z = 0 begins to rotate in counter-clockwise direction while propagating in the z-direction (out of the page). Using our right hand, we can align the thumb in the propagation direction (out of the page), and the fingers curve around (counter-clockwise) to represent the field rotation. Therefore, this is called the Right-Handed Circular Polarization (RHCP), and is described by Eqn.(6.9) – two linear wave differ by 90-degrees in space and 90-degrees in phase.

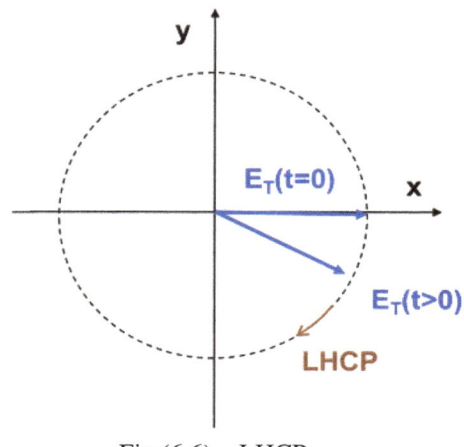

Fig.(6.6) – LHCP

Likewise, the equations to describe the Left-Handed Circular Polarization (LHCP) can be written as:

$$\vec{E}_{LHCP}(\vec{r},t) = \hat{x}E_o \cos(\omega t - \beta z) - \hat{y}E_o \sin(\omega t - \beta z)$$
$$\vec{E}_{LHCP}(\vec{r},t) = \hat{x}E_o e^{j(\omega t - \beta z)} - \hat{y}E_o e^{j(\omega t - \beta z - \pi/2)}$$
$$\vec{E}_{LHCP}(\vec{r},t) = (\hat{x} + j\hat{y})E_o e^{j(\omega t - \beta z)} \qquad \text{Eqn.(6.10)}$$

Again, when the 2 linearly polarized waves are 90-degrees apart in phase, and 90-degrees apart in space (spatially), they form a circularly polarized wave if and only if they have the same amplitude as well.

Example (6.6) - $\vec{E}(\vec{r},t) = \hat{x}5\sin(10^8 t + 2y) - \hat{z}5\cos(10^8 t + 2y)$ V/m. What is the polarization of this wave?

Let us call the cosine term E_1, and the sine term E_2.
At time = 0 and at origin, E_1 is along the –z direction.
As time goes on, the resultant E-field is rotating counter-clockwise, according to the axes shown in the figure.
Propagation (k) direction is out of the page.
So this is a RHCP.

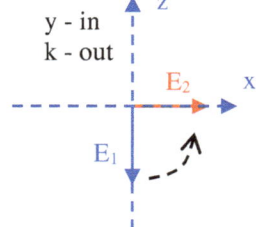

Example (6.7) - A LHCP is propagating in the direction of $\hat{k} = \dfrac{3\hat{x}-4\hat{y}}{5}$.
At the origin when t = 0, E = 2 V/m along the z-axis.
Write the E-field.

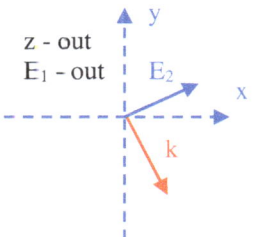

Using the left hand, E_2 is found to be in the 1st quadrant when E_1 is pointing out of the page and the thumb is along the k direction.

$$\vec{E}(\vec{r},t) = \hat{z}2\cos\left(\omega t - \dfrac{k}{5}(3\hat{x}-4\hat{y})\right) + 2\left(\dfrac{4\hat{x}+3\hat{y}}{5}\right)\sin\left(\omega t - \dfrac{k}{5}(3\hat{x}-4\hat{y})\right) \text{ V/m}$$

Many creatures in nature display or utilize circularly polarized light. The Rose Chafers (a type of beetle only live through spring – summer time) reflect a LHCP green light from their shells primarily due to the helical molecular structure in the shells.

The Mantis Shrimps, also known as the "pissing shrimp" (攋尿蝦) in Asia, are known for their ablility to sense LHCP and RHCP light over a wide spectrum of color. They have perhaps the most complex eye structure in the animal kingdom. Recent studies showed that they use the circularly polarized light for communications among themselves.

6.5.3 Elliptical polarization

If the amplitude of the 2 perpendicular linearly polarized waves are not the same, the resultant wave would sweep in a circularly fashion but with a major and a minor axis. In other words, we have an elliptically polarized wave. For example:

$$\vec{E}(\vec{r},t) = \hat{x}2\cos(\omega t - \beta z) + \hat{y}3E_o \sin(\omega t - \beta z)$$

is a RHEP (Right-Handed Elliptical Polarization), with y-axis is the major axis. In general, it can be expressed as:

$$\vec{E}(\vec{r},t) = \hat{x}E_{o1}\cos(\omega t - \beta z) \pm \hat{y}E_{o2}\sin(\omega t - \beta z)$$
$$\vec{E}(\vec{r},t) = (\hat{x}E_{o1} \mp j\hat{y}E_{o2})e^{j(\omega t - \beta z)}$$

Eqn.(6.11)

where the top sign is the RHEP, and the bottom sign is LHEP.

Another way to get an elliptical polarization is when the phases are not exact 90 degrees apart, even though the amplitudes of the 2 linearly polarized waves are the same. For example: $\vec{E}(\vec{r},t) = \hat{x}2\cos(\omega t - \beta z) + \hat{y}2E_o\sin(\omega t - \beta z + 0.1)$ is a RHEP.

In short, to get a circularly polarized wave, both amplitudes have to be the same, and the phases must be 90 degrees apart.

$$\vec{E}(\vec{r},t) = \hat{x}E_{o1}\cos(\omega t - \beta z) \mp \hat{y}E_{o2}\cos(\omega t - \beta z + \delta)$$
$$\vec{E}(\vec{r},t) = \hat{x}E_{o1}e^{j(\omega t - \beta z)} \mp \hat{y}E_{o2}e^{j(\omega t - \beta z + \delta)}$$
$$\vec{E}(\vec{r},t) = (\hat{x}E_{o1} \mp \hat{y}E_{o2}e^{j\delta})e^{j(\omega t - \beta z)}$$

Eqn.(6.12)

Equations (6.12) express the generic form of elliptical polarization assuming the wave is propagating in the z-direction. If δ = 90 degrees and $E_{o1} = E_{o2}$, then we have a circular polarization.

<u>Example (6.8)</u> - $\vec{E}(z,t) = \hat{x}3\cos(\omega t - kz + \pi/6) - \hat{y}4\sin(\omega t - kz + \pi/4)$

What is polarization of the wave?

$E_{o1} \neq E_{o2}$ and phases are not the same → elliptical. Although the phase is not totally correct, but the sense of left-handed is demonstrated.

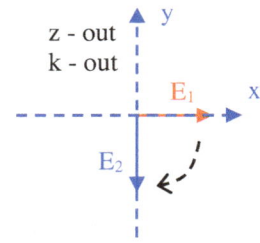

To be more precise, we can plot (for z = 0) the E_1 and the E_2 components in a spreadsheet.

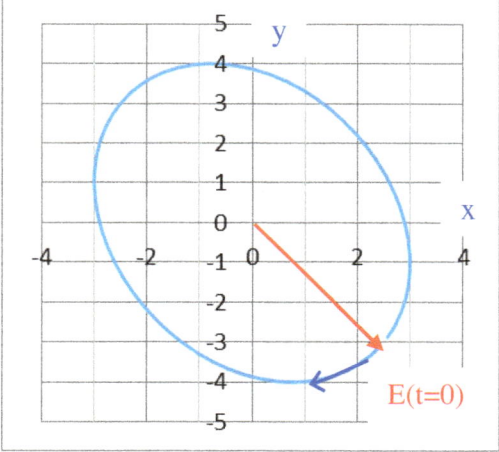

The resultant electric field is rotating in the clockwise direction as shown, and the wave is propagating outward in the z-direction. It is clearly an elliptical polarization. The major and minor axes are tilted. There are mathematical formulas to calculate the tilted angle. In practice, however, this phase shift can be adjusted by relocating the receive antenna during calibration.

As mentioned earlier, any electromagnetic wave can be written as a linear combination of a horizontally polarized and a vertically polarized wave. It can also be written as a linear combination of a right-handed and a left-handed circularly polarized wave. This is familiar in saying that any vector can be expressed in Cartesian coordinates using unit vectors in x and y directions, or in polar coordinates using unit vectors in r and θ directions. In fact, any 2 unit vectors that can span the space can be used as the basis vectors. Therefore, any electromagnetic wave can be written as a linear combination of the left-handed and right-handed elliptically polarized wave as well.

Example (6.9) - Write $\vec{E}(z,t) = \hat{x} E_o \cos(\omega t - kz)$ as a combination of RHCP and LHCP.

$$\vec{E}_{RHCP}(\vec{r},t) = (\hat{x} - j\hat{y}) E_1 e^{j(\omega t - \beta z)}$$
$$\vec{E}_{LHCP}(\vec{r},t) = (\hat{x} + j\hat{y}) E_2 e^{j(\omega t - \beta z)}$$
$$\vec{E}_T(\vec{r},t) = \vec{E}_{RHCP}(\vec{r},t) + \vec{E}_{LHCP}(\vec{r},t)$$
$$\vec{E}_T(\vec{r},t) = [\hat{x}(E_1 + E_2) + j\hat{y}(E_2 - E_1)] e^{j(\omega t - \beta z)}$$
$$E_2 = E_1 = \frac{E_o}{2}$$

More explicitly,

$$\vec{E}_{RHCP}(z,t) = \hat{x}\frac{E_o}{2}\cos(\omega t - kz) + \hat{y}\frac{E_o}{2}\sin(\omega t - kz)$$
$$\vec{E}_{LHCP}(z,t) = \hat{x}\frac{E_o}{2}\cos(\omega t - kz) - \hat{y}\frac{E_o}{2}\sin(\omega t - kz)$$
$$\vec{E}_{linear} = \vec{E}_{RHCP}(z,t) - \vec{E}_{LHCP}(z,t)$$

6.6   Polarizer

A polarizer is an optical filter that eliminate all light except those with the electric field lines up with the optical axis defined by its crystal structure. If the incoming light is unpolarized, the electromagnetic wave would be polarized after passing through the polarizer and the power intensity would be decreased by half. If the incoming light is linearly polarized, the exiting wave would be proportional to $\cos^2\theta$ where θ is defined as the angle between the optical axis and the incoming electric field vector.

Fig.(6.7) – An unpolarized light go through 2 polarizers with an angle θ between their optical axes.

unpolarized, I_o intensity; polarized, $I_1 = ½ I_o$; polarized, $I_2 = I_1 \cos^2\theta = ½ I_o \cos^2\theta$

If the second polarizer is rotated 90 degrees with respect to the first one, then no light can go through ($\cos 90° = 0$).

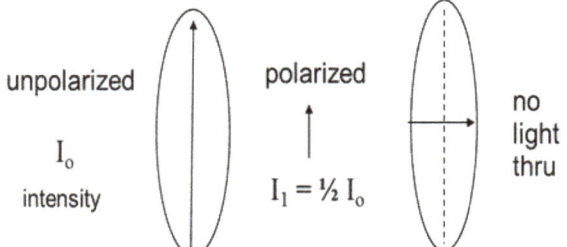

Fig.(6.8) – An unpolarized light go through 2 polarizers with 90 degrees between their optical axes.

However, if a third polarizer is placed in between the 2 perpendicularly polarized lens in Fig.(6.8), then a polarized wave would go through!

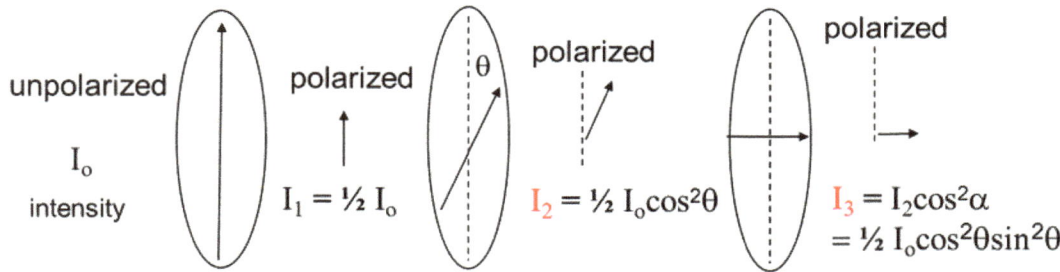

Fig.(6.9) – An unpolarized light go through 3 polarizers with the first and last one are 90 degrees between their optical axes.

Some material with complex molecular structures, such as helical structure, can rotate the electric field and allow light to go through. The rotational axis sometimes depends on the wavelength or color as shown in Fig.(6.11). Some crystal rotates the field only when a voltage is applied to it. Such feature is used in calculators or handset display applications.

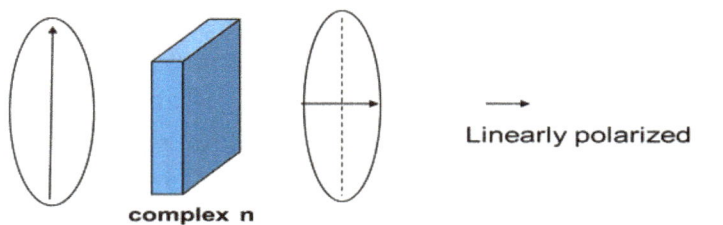

Fig.(6.10) – A structure with complex index of refraction is placed in between 2 polarizers.

Many simple materials, such as some adhesive tape, sugar solution, and a certain type of plastic. Below is some overlaid tape placed between a pair of polarizers rotated at 90 degrees with respect to each other. Different thickness of tape rotate different color light at different rates (wavelength dependent). And after the rotated light passes through the second polarizer, only a certain color stands out.

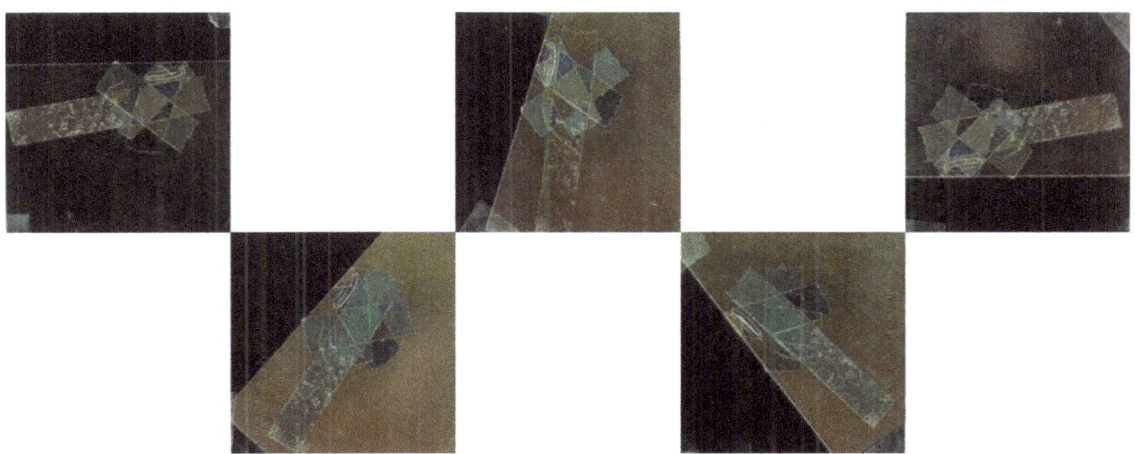

Fig.(6.11) – Layers of adhesive tape rotated in between 2 orthogonal polarizers. The emerged color depends on the orientation of both polarizers, as well as the thickness of the tape.

More color effect can be obtained by rotating the polarizers with respect to each other indeed. The combinations and possibilities are endless.

Fig.(6.12) – Layers of adhesive tape placed between 2 polarizers. Keeping the tape and the first polarizer stationary, this set of color pattern is obtained by rotating the second polarizer only.

Exercise:

6.1 Given that $\vec{E}(\vec{r},t) = \hat{y}0.1\sin(10\pi x)\cos(6\pi \cdot 10^9 t - \beta z)$ V/m in air, find H(r,t) and β using the phasor equations.

6.2 A 60 MHz electromagnetic wave exists in an air-dielectric coaxial cable having an inner conductor with radius a and an outer conductor with inner radius b. Assuming perfect conductors, and the phasor form of the electric field intensity to be (a < r < b)
$$\vec{E} = \hat{r}\frac{E_o}{r}e^{-jkz} \text{ V/m},$$
(a) find k,
(b) find H from the $\nabla \times \vec{E} = -j\omega\mu\vec{H}$,
(c) find the surface current densities on the inner and outer conductors.

6.3 A 159.2 MHz LHCP electromagnetic wave is propagating in the "–y" direction. The electric field is found to be 5 V/m along the z-axis at the origin when t = 0. What is the corresponding H field of the wave?

6.4 The instantaneous expression for the magnetic field intensity of a uniform plane wave propagating in the +y direction in air is given by $\vec{H} = \hat{z}4\cdot 10^{-6}\cos(10^7\pi t - ky + \pi/4)$ A/m,
(a) Determine k and the location where H_z vanishes at t = 3 ms,
(b) Write the instantaneous expression for **E**.

6.5 The E-field of a uniform plane wave propagating in a dielectric medium is given by
$$\vec{E} = \hat{x}2\cos(10^8 t - z/\sqrt{3}) - \hat{y}\sin(10^8 t - z/\sqrt{3}) \text{ V/m},$$
(a) Determine frequency and wavelength of the wave,
(b) What is the dielectric constant of the medium?
(c) Describe the polarization of the wave.
(d) Find the corresponding H-field.

7 Normal incidence

In the previous chapters, we learned that electromagnetic wave is generated whenever the electromagnetic fields are varying over time, such as from a AC source. We also learned how to describe the electromagnetic wave travelling in a medium. Next, we are going to examine the wave behavior when it encounters a change in medium. In this chapter, we consider only the case of normal incidence, i.e., the wave propagation direction is perpendicular to the boundary of the interface.

7.1 The wave equations

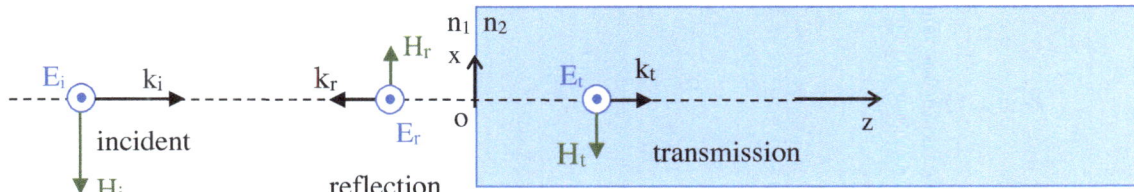

Fig.(7.1) – A normal incidence.

Figure (7.1) represents an electromagnetic wave, incident perpendicularly to the interface defined by the change of medium (index of refraction) from n_1 to n_2. Some of the wave would be reflected and traveling in the reverse direction as indicated by k_r in the figure. The rest of the wave would continue to enter the new medium. Assuming the media are homogeneous, isotropic, and linear dielectrics, the electric field equations can be written, without lost of generality, as:

$$\vec{E}_i(z,t) = \hat{y}E_{oi}e^{j(\omega_i t - k_i z)}$$
$$\vec{E}_r(z,t) = \hat{y}E_{or}e^{j(\omega_r t + k_r z)}$$
$$\vec{E}_t(z,t) = \hat{y}E_{ot}e^{j(\omega_t t - k_t z)} \qquad \text{Eqn.(7.1)}$$

The wave amplitudes can be positive or negative, meaning that we do not assume all electric fields are in phase at this point. In fact, we do not even assume the frequencies are the same.

Applying the boundary condition of Eqn.(5.21) that tangential component of the electric field to be continuous across the boundary to the origin in Fig.(7.1), we have:

$$E_{oi}e^{j\omega_i t} = E_{or}e^{j\omega_r t} = E_{ot}e^{j\omega_t t} \qquad \text{Eqn.(7.2)}$$

which is true for any time t if and only if $\omega_i = \omega_r = \omega_t$. In other words, the frequency does not change upon reflection or crossing a boundary to a homogeneous and isotropic medium.

The wave number $k = 2\pi/\lambda$ only depends on the medium. So $|k_i| = |k_r| = |k_1|$.

$$|k_i| = |k_r| = k_1 = \frac{\omega}{v_1} = \frac{\omega n_1}{c} = k_o n_1$$

$$|k_t| = k_2 = k_o n_2 \qquad \text{Eqn.(7.3)}$$

The accompany H-field can be written, along with the E-field, as:

$$\vec{E}_i(z,t) = \hat{y} E_{oi} e^{j(\omega t - k_1 z)}$$

$$\vec{H}_i(z,t) = -\hat{x} H_{oi} e^{j(\omega t - k_1 z)} = -\hat{x} \frac{E_{oi}}{\eta_1} e^{j(\omega t - k_1 z)}$$

$$\vec{E}_r(z,t) = \hat{y} E_{or} e^{j(\omega t + k_1 z)}$$

$$\vec{H}_r(z,t) = \hat{x} H_{or} e^{j(\omega t + k_1 z)} = \hat{x} \frac{E_{or}}{\eta_1} e^{j(\omega t + k_1 z)}$$

$$\vec{E}_t(z,t) = \hat{y} E_{ot} e^{j(\omega t - k_2 z)}$$

$$\vec{H}_t(z,t) = -\hat{x} H_{ot} e^{j(\omega t - k_2 z)} = -\hat{x} \frac{E_{ot}}{\eta_2} e^{j(\omega t - k_2 z)} \qquad \text{Eqn.(7.4)}$$

Apply the boundary condition Eqn.(5.21) again to the electric field equations, we have:

$$E_{oi} + E_{or} = E_{ot} \qquad \text{Eqn.(7.5)}$$

Similarly, the boundary condition for the tangential H-field between 2 dielectric medium where there is no surface current is simply: $H_{t1} = H_{t2}$. Hence we get another expression from Eqn.(7.4):

$$H_{oi} - H_{or} = H_{ot}$$

$$\frac{E_{oi} - E_{or}}{\eta_1} = \frac{E_{ot}}{\eta_2} \qquad \text{Eqn.(7.6)}$$

Combining Equations (7.5) and (7.6), we can solve for E_{or} and E_{ot}:

$$\eta_1(E_{oi} + E_{or}) = \eta_1(E_{ot})$$

$$\eta_2(E_{oi} - E_{or}) = \eta_1(E_{ot})$$

$$\frac{E_{or}}{E_{oi}} = \frac{\eta_2 - \eta_1}{\eta_2 + \eta_1} \equiv \Gamma \qquad \text{Eqn.(7.7)}$$

$$\eta_2(E_{oi} + E_{or}) = \eta_2(E_{ot})$$

$$\eta_2(E_{oi} - E_{or}) = \eta_1(E_{ot})$$

$$\frac{E_{ot}}{E_{oi}} = \frac{2\eta_2}{\eta_1 + \eta_2} \equiv \tau = 1 + \Gamma \qquad \text{Eqn.(7.8)}$$

Γ in Eqn.(7.7) is called the Reflection Coefficient. It represents the fraction of electric field being reflected at the boundary. And τ in Eqn.(7.8) is the Transmission Coefficient. It is the fraction of the electric field being transmitted into the second medium. So given the incident wave and known medium, the reflected and transmitted waves can be predicted precisely.

Example (7.1) - A 159.2 MHz signal is propagating along the x-axis in air, incidented normally onto a dielectric of n = 1.5, with a 20 V/m electric field in the z-direction at the origin when t = 0, write the explicit time function of all the electromagnetic fields.

$$\omega = 2\pi f = 2\pi(159.2 \cdot 10^6) = 10^9 [rad/s]$$

$$k_1 = k_o = \frac{\omega}{c} = \frac{10^9}{3 \cdot 10^8} = \frac{10}{3} = 3.33 [m^{-1}]$$

$$\vec{E}_i = \hat{z} 20 \cos(10^9 t - 3.33x) [V/m]$$

$$\vec{H}_i = -\hat{y} \frac{20}{\eta_o} \cos(10^9 t - 3.33x) = -\hat{y} 53 \cos(10^9 t - 3.33x) [mA/m]$$

$$n_2 = 1.5 = \sqrt{\varepsilon_{r2}}$$

$$\varepsilon_{r2} = 1.5^2 = 2.25$$

$$\eta_2 = \sqrt{\frac{\mu_2}{\varepsilon_2}} = \eta_o \sqrt{\frac{\mu_{r2}}{\varepsilon_{r2}}} = \eta_o \sqrt{\frac{1}{\varepsilon_{r2}}} = \frac{\eta_o}{n_2}$$

$$\Gamma = \frac{\eta_2 - \eta_1}{\eta_2 + \eta_1} = \frac{\frac{\eta_o}{n_2} - \frac{\eta_o}{n_1}}{\frac{\eta_o}{n_2} - \frac{\eta_o}{n_1}} = \frac{n_1 - n_2}{n_1 + n_2} = \frac{1 - 1.5}{1 + 1.5} = -\frac{1}{5} = -0.2 = \frac{E_{or}}{E_{oi}}$$

$$E_{or} = -0.2 E_{oi} = -0.2(20) = -4 [V/m]$$

$$H_{or} = \frac{E_{or}}{\eta_o} = \frac{4}{377} = 10.6 [mA/m]$$

$$\vec{E}_r = -\hat{z} 4 \cos(10^9 t + 3.33x) [V/m]$$

$$\vec{H}_r = \hat{y} 10.6 \cos(10^9 t + 3.33x) [mA/m]$$

$$\tau = 1 + \Gamma = 1 - 0.2 = 0.8 = \frac{E_{ot}}{E_{oi}}$$

$$E_{ot} = 0.8 E_{oi} = 0.8(20) = 16 [V/m]$$

$$H_{ot} = \frac{E_{ot}}{\eta_2} = \frac{E_{ot}}{\eta_o} n_2 = \frac{16}{377}(1.5) = 63.7 [mA/m]$$

$$k_2 = k_o n_2 = 3.33(1.5) = 5 \left[m^{-1} \right]$$
$$\vec{E}_t = \hat{z} 16 \cos\left(10^9 t - 5x\right) [V/m]$$
$$\vec{H}_t = -\hat{y} 63.7 \cos\left(10^9 t - 5x\right) [mA/m]$$

So, all 6 electromagnetic field equations are calculated. A few useful observations in this example are worth pointing out.

1. Electric field is expressed in Volt per meter [V/m], whereas H-field is often in milli-amp per meter [mA/M].
2. The wave impedance η = η$_o$ / n is true for non-magnetic material, which is what we are dealing with in wave propagation unless we are explicitly working in ferrite devices.
3. Γ is negative here only means the reflected electric field is 180 degrees out-of-phase with respect to the incident wave.
4. $\Gamma = \dfrac{\eta_2 - \eta_1}{\eta_2 + \eta_1} = \dfrac{n_1 - n_2}{n_1 + n_2}$ is always valid for non-magnetic material.

   It was first introduced in wave optics when dealing with thin film interference problems. Light reflected from a surface is in-phase with the incident light if n$_1$ > n$_2$ (such as from water to air), and 180 degrees out-of-phase if n$_1$ < n$_2$ (such as from air to glass).

5. H$_{or}$ is written as a positive number even though E$_{or}$ < 0. The negative direction of E$_r$ is taken care of in the diagram, so to avoid double counting. Readers are free to handle the negative sign in other fashions, but the final answer has to be consistent.

6. The transmission coefficient τ = 1 + Γ is true for the case of normal incidence and other polarization. It is NOT a general statement and should be cautious when applying this formula.

7. The reflection coefficient Γ = (η2 − η1)/(η2 + η1) is a general statement which keeps appearing and worth memorizing. The concept of "the difference in impedance divide by the sum" will be discussed in later chapters. Simply put, part of the incident wave will be reflected if there is a different in wave impedance in the medium, regardless whether the impedance is higher or lower. Same is true for water wave, for example. Some water wave will be reflected whenever there is a discontinuity in depth regardless the floor steps up or down.

7.2 Reflection coefficient

$$\Gamma = \frac{\eta_2 - \eta_1}{\eta_2 + \eta_1} = \frac{n_1 - n_2}{n_1 + n_2} \qquad \text{Eqn.(7.9)}$$

As in Example (7.1), the wave going from a lower index of refraction (air) to a higher index medium (glass) would result a negative Γ, which means the reflected wave carries a 180-degrees phase shift with respect to the incident wave.

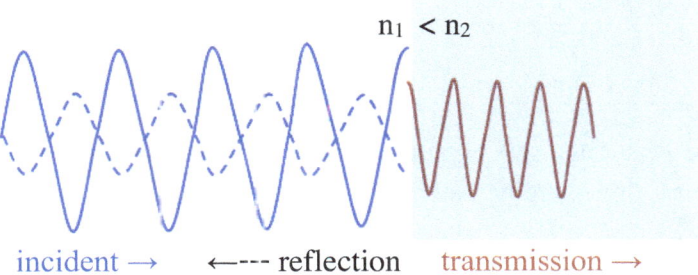

Fig.(7.2) – Electric field. Normal incident, from lower to higher index of refraction.

Material with a lower index of refraction often has a lower mass density as well. It is sometimes useful to visualize the situation. For example, the wave is traveling from air to glass in Figure (7.2), so it signifies a transition from a lower index to a higher index of refraction. One can see that the incidence wave and the reflected wave have the same wavelength because they are in the same medium. The reflected wave is 180-degrees out-of-phase with respect to the incidence wave. The transmitted wave is in-phase with respect to the incidence. The wavelength in the second medium is shorter because of its higher index of refraction (denser).

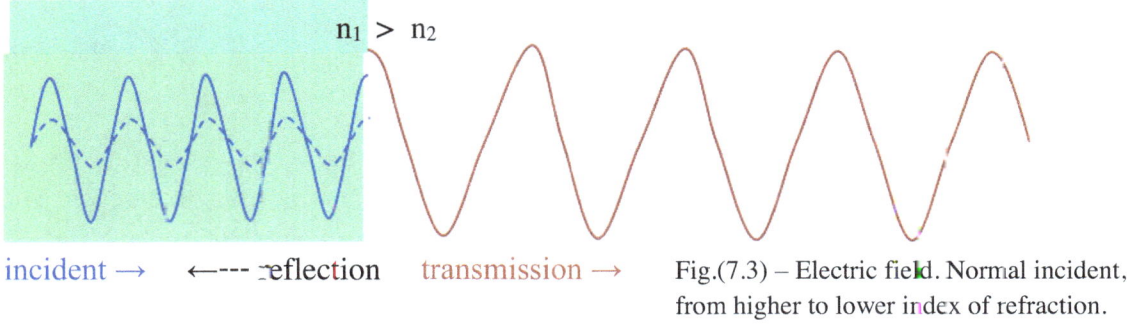

Fig.(7.3) – Electric field. Normal incident, from higher to lower index of refraction.

Figure (7.3) depicts the other scenario where the wave is incident from a denser medium (water) to a lower index medium (air). The incident, reflected and transmitted waves are all in phase with each other. The wavelength in medium 2 is longer because the index of refraction is lower (less dense). The amplitude of the transmitted wave is also larger than the incident wave! Does that violate conservation of energy? Of course it does not. The energy of the wave is not only proportional to the $|E|^2$, but $(½)\varepsilon|E|^2$. Lower index of refraction also means a smaller ε_r. The total incident energy has to be equal to the sum of the reflected and transmitted energy. The electromagnetic radiation power will be discussed in the next section.

7.3 Power consideration

The electric energy density and power density (Poynting vector) are given by:

$$U_E = \frac{1}{2}\varepsilon |E|^2$$

$$S_{ave} = \frac{1}{2}\frac{|E|^2}{\eta}$$

Eqn.(7.10)

Since the reflected wave and incident wave are in the same medium with the same wave impedance, the power ratio can simply be written as:

$$\frac{P_r}{P_i} = \left|\frac{E_{or}}{E_{oi}}\right|^2 = |\Gamma|^2$$

Eqn.(7.11)

The transmitted wave and the incident wave are in different medium, so the wave impedance has to be kept in the expression:

$$\frac{P_t}{P_i} = \frac{\eta_1}{\eta_2}\left|\frac{E_{ot}}{E_{oi}}\right|^2 = \frac{\eta_1}{\eta_2}|\tau|^2$$

Eqn.(7.12)

But conservation of energy dictates that $P_i = P_r + P_t$. So,

$$1 = \frac{P_r}{P_i} + \frac{P_t}{P_i} = |\Gamma|^2 + \frac{P_t}{P_i}$$

$$\frac{P_t}{P_i} = 1 - |\Gamma|^2$$

Eqn.(7.13)

Equation (7.11) is a very general statement and most useful in calculating the fractional power being reflected. Equation (7.12) is true for normal incidence and some polarization only, whereas Eqn.(7.13) comes from the conversation law and is always valid. Therefore, Eqn.(7.13) should be the preferred formula whenever possible.

7.4 Standing Wave Ratio (SWR)

The incident wave and reflected wave in Figures (7.2) and (7.3) are 2 waves with the same wavelength traveling in opposite directions. This is how standing waves are formed. (See Appendix E for a quick review.) The maximum amplitude of this standing wave is sum of the amplitudes of the incident wave and the reflected wave. And the minimum amplitude of this standing wave is the difference of the two amplitudes. The resultant standing wave pattern is clearly illustrated in Appendix E. The ratio of the maximum amplitude to the minimum amplitude is called the Standing Wave Ratio (SWR).

$$SWR \equiv \frac{E_{max}}{E_{min}} = \frac{E_{incident} + E_{reflected}}{E_{incident} - E_{reflected}} = \frac{E_1 + E_2}{E_1 - E_2} \qquad \text{Eqn.(7.14)}$$

SWR ranges from 1.0 (indicating zero reflection) to infinity (in the case of total reflection). It is a parameter used to indicate how different the materials are, or equivalently, how different the impedance of the waves are across the boundary.

<u>Example (7.2)</u> - A plane wave in air with $\vec{H}(\vec{r},t) = \hat{x}2\sin(\omega t - 4z)$ mA/m, is incident upon the planar surface of a dielectric material with n = 1.5, occupying the half-space z > 0. Write the full expression for the reflected E field and the transmitted H field. What is the reflected power and transmitted power in dB?

$$\omega = ck = (3 \cdot 10^8)(4) = 1.2 \cdot 10^9$$
$$\vec{H}_i = \hat{x}2\sin(\omega t - 4z)$$
$$\vec{E}_i = (-\hat{y})(2\eta_o)\sin(\omega t - 4z)$$
$$\eta_2 = \frac{\eta_0}{n_2} = \frac{377}{1.5} = 251$$
$$\Gamma = \frac{\eta_2 - \eta_0}{\eta_2 + \eta_0} = \frac{251 - 377}{251 - 377} = -0.2$$
$$\vec{E}_r(\vec{r},t) = (-0.2)(-\hat{y})(2\eta_o)\sin(\omega t + 4z) = \hat{y}151\sin(1.2 \cdot 10^9 t + 4z) \quad \text{mV/m}$$

$$\tau = 1 + \Gamma = 1 - 0.2 = 0.8$$
$$k_2 = \frac{\omega}{v_2} = \frac{\omega n_2}{c} = k_o n_2 = (4)(1.5) = 6$$
$$\vec{E}_t(\vec{r},t) = (0.8)(-\hat{y})(2\eta_o)\sin(\omega t - k_2 z) = -\hat{y}1.6\eta_o \sin(\omega t - 6z)$$
$$\vec{H}_t(\vec{r},t) = \hat{x}\frac{1.6\eta_o}{\eta_2}\sin(\omega t - 6z) = \hat{x}1.6n_2 \sin(\omega t - 6z) = \hat{x}2.4\sin(\omega t - 6z)$$

$$\frac{P_r}{P_i} = |\Gamma|^2 = 0.04 = 4\% = 20\log|\Gamma| = 20\log|-0.2| = -14 dB$$

mA/m

$$\frac{P_t}{P_i} = 1 - |\Gamma|^2 = 0.96 = 96\% = 10\log\left|1 - |\Gamma|^2\right| = 10\log|0.96| = -0.18 dB$$

Example (7.3) - A 159 MHz RHCP electromagnetic wave is incident normally (along the z-direction) from air onto a lossless dielectric medium ($\varepsilon_r = 4$, $z > 0$) such that $E_{oi}(0,0)$ at the origin is 2 V/m along the x-direction. Write the incident $\mathbf{E}_i(r,t)$ and reflected $\mathbf{E}_r(r,t)$ wave expressions. What is the polarization state of the reflected wave? What is the reflected power and transmitted power in dB?

$$\vec{k}_i = k_o \hat{z}$$

$$\vec{E}_i(\vec{r},t) = \hat{x} 2\cos(\omega t - k_o z) + \hat{y} 2\sin(\omega t - k_o z)$$

$$\omega = 2\pi f = 10^9$$

$$k_o = \frac{\omega}{c} = \frac{10}{3} = 3.33$$

$$\vec{E}_i(\vec{r},t) = \hat{x} 2\cos(10^9 t - 3.33z) + \hat{y} 2\sin(10^9 t - 3.33z) \text{ V/m}$$

$$\Gamma = \frac{\eta_2 - \eta_1}{\eta_2 + \eta_1} = \frac{n_1 - n_2}{n_1 + n_2} = \frac{1-2}{1+2} = -\frac{1}{3}$$

$$\vec{k}_r = -k_o \hat{z} = -3.33\hat{z}$$

$$\vec{E}_r(\vec{r},t) = -\hat{x}\frac{2}{3}\cos(10^9 t + 3.33z) - \hat{y}\frac{2}{3}\sin(10^9 t + 3.33z) \text{ V/m}$$

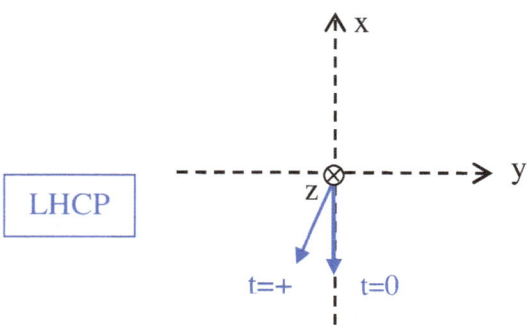

LHCP

$$\frac{P_r}{P_i} = |\Gamma|^2 = \left|-\frac{1}{3}\right|^2 = \frac{1}{9} = 0.111 = 11.1\% = 20\log|\Gamma| = 20\log|-0.33| = -9.5 dB$$

$$\frac{P_t}{P_i} = 1 - |\Gamma|^2 = \frac{8}{9} = 88.9\% = 10\log|1 - |\Gamma|^2| = 10\log|0.889| = -0.5 dB$$

8 Transmission Line Theory

A wireless signal traveling from a cell tower to our handset, for example, can be adequately described using Maxwell's equations. One can imagine how the electromagnetic field oscillates in space or in a medium from one location to another. But how does this signal travel in our cell phone through all the electronic devices? The electromagnetic field is far too complex to visualize. Early pioneers of microwave and RF (Radio Frequency) engineers developed equivalent circuits of many structures to simplify the task.

8.1 RF circuit representation

A microwave or RF circuit is different than the low frequency counterpart in many aspects. Grounding is an example. Its construction is just as important, if not more, than the signal path. A single point of contact is sufficient for dc ground or low frequency application. RF ground is far more important and need a lot of attention. Traditionally, the grounding path is explicitly included in the circuit symbols. Even a one-port network, such as an antenna, or a connector, where there is only 1 connection to be made, it has to have 2 terminals or wires as shown in Fig.(8.1) below.

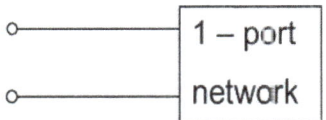

Fig.(8.1) – A 1-port network is represented with 2 terminals in a RF circuit.

For a 2-port network, there will be 4 terminals or wires. Conventionally, the input port is on the left and the output port is on the right, although one can re-arrange it in any other ways. The input port is usually connected to the source: the power source or signal source. The output end is connected to the load, which can represent another device or a chain of devices, or simply a resistor. It is useful to visualize a RF circuit as a flow of energy, or signal power, from the source to the load. After all, it is the flow of electromagnetic wave we want to utilize.

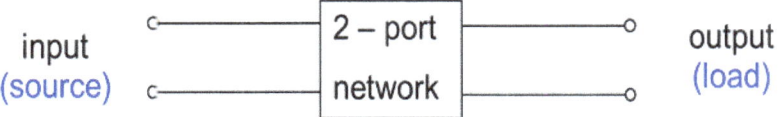

Fig (8.2) – A 2-port network is represented with 4 terminals in a RF circuit. Signal flows from left to right.

Circuit elements can be connected together in series, in which they share the same electrical current, or connected in parallel where they share the same voltage. The schematic of the RF connections look somewhat different.

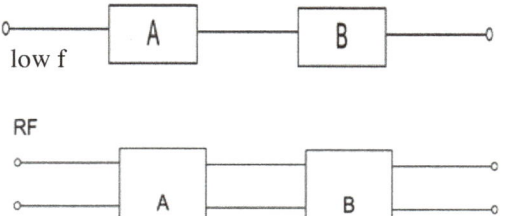

Fig.(8.3) – A series connection of 2 circuit elements in low frequency schematic, and in RF representation.

Again, the RF circuit assumes the source is on the left and the load is on the right. The relation of the source to the circuits is explicitly sketched out in Fig.(8.4).

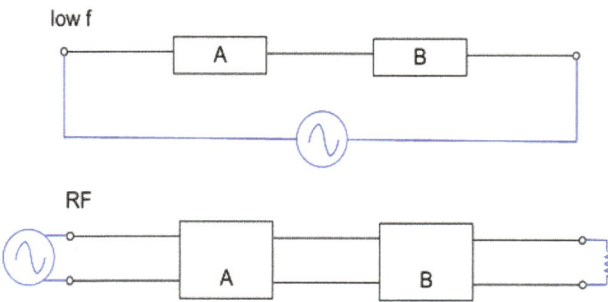

Fig.(8.4) – A series connection of 2 circuit elements in a typical low frequency schematic, and in a RF circuit.

Likewise, a parallel connection is also different between the RF circuit and its low frequency counterpart.

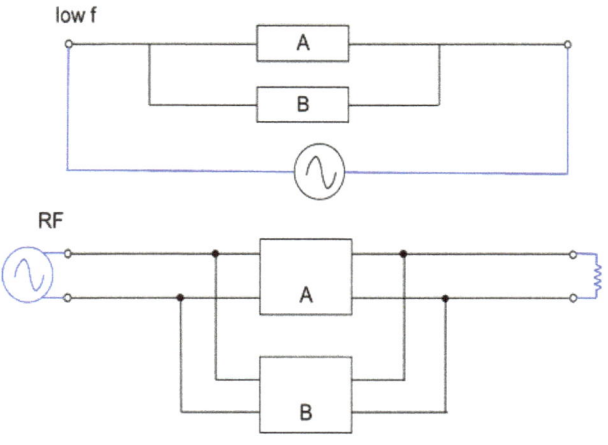

Fig.(8.5) – A parallel connection of 2 circuit elements in a typical low frequency schematic, and in a RF circuit.

Of course, elements A and B are generic elements which can be as simple as a resistor, or a complex network of components. More detailed schematic will be discussed later in the chapter.

8.2 Transmission lines

A transmission line is structure that guides the signal, or electromagnetic wave, to travel in the desired direction. Below is a list of common transmission lines people use in communication systems.

8.2.1 Twisted pair

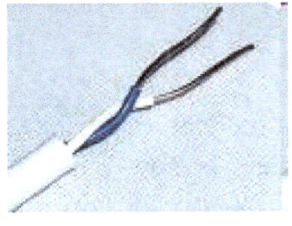

This is one of the early attempts for higher frequency transmission. It is typically used in VLF (Very Low Frequency) range below 30 kHz. The twisted wires minimize the area between the wires so that any noise induced due to magnetic fluctuation is minimized. More importantly, the direction of the induced current would be alternated and cancel out.

8.2.2 Parallel wires

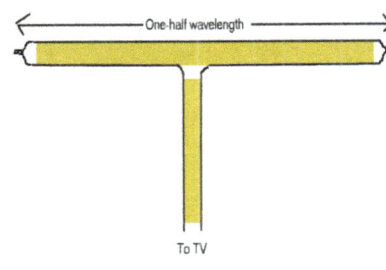

As frequency increases, the twisting direction and somewhat varying separation of the wires induce significant and unnecessary loss. At LF (Low Frequency below 300 kHz) to HF (High Frequency below 30 MHz), some applications use parallel wires. The 2 wires are separated by a plastic (nylon) and the impedance become very steady. This was typically used in old radio receivers as antenna.

8.2.3 Coaxial cable

This is one of the most common transmission lines used commercially. It is very flexible, and covers a wide range of frequency without distortion. The cost is relatively low unless it is made for special applications in very high frequency or ultra low loss. The geometry is simple so the characteristic of the cable can be calculated precisely. The electromagnetic wave is propagating between the center conductor and the outer conducting shield. If the dielectric between the conductors is low loss, coaxial cable can be a high performance transmission line.

8.2.4 Microstrip line

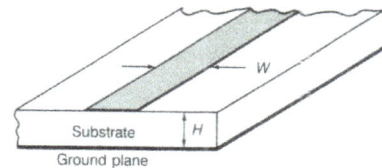

Microstip line is substrate based transmission line similar to a pc board. The top trace is etched or printed on the top surface, and the bottom surface is usually a layer of copper for ground plane.

The signal propagates through the substrate between the conductors. Depends on the frequency of operation, specific substrate material is selected for its loss, cost, manufacturability, and availability.

Microstrip is by far the most common transmission line in commercially applications because of its low manufacturing cost and relatively easy fabrication. It is also very adaptable to integrate with other electronic devices. Similar to the coaxial cable, microstrip line also offers a wide frequency range without frequency distortion.

8.2.5 Coplanar waveguide (CPW)

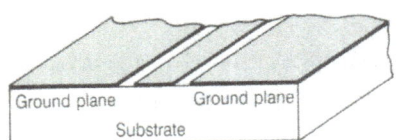

Coplanar waveguide is very similar to the microstrip line except the bottom ground plane is now put on the top surface as well. The bottom surface is left non-conducting. Mathematically, it is treated similar to the microstrip with conformal mapping, meaning the space between the top trace and the bottom ground plane in microstrip is now curved to the sides.

It enjoys the same benefit of low manufacturing cost and wide operating frequency as the microstrip line. The biggest advantage of the coplanar structure is its convenience to design in with surface mount chipsets. All dropin chipsets require ground pins connection. Any via holes or other structural connections to ground introduce unnecessarily parasitic fringing which can alter the performance of the design. CPW provides an easy contact with minimal parasitic penalty.

Similar to the coaxial cable and microstrip line, CPW operates on a TEM mode of wave propagation (to be explained later). So strictly speaking, it is not a waveguide in the usual sense. The reason it is called the coplanar waveguide was because the inventor C.P. Wen has the same initial CPW.

8.2.6 Waveguide

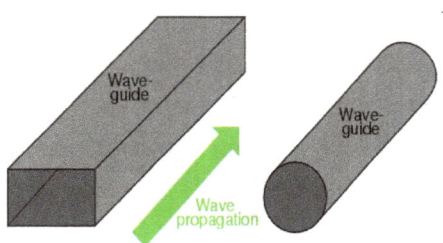

Waveguides are hollow conducting pipes, usually made of copper tubes, for its high elcetrical conductivity. They are the highest performance transmission lines in terms of transmission loss and power handling capability. However, they are very frequency limited (divided in bands) and usually rigid in construction.

Waveguide is perhaps best described using electromagnetic fields and Maxwell's equations in understanding the characteristic of signal transmission. Nonetheless, equivalent circuits do exist and used widely especially for structural changes within the waveguide for frequency selection, phase variation or polarization alternation.

8.3 Equivalent circuit

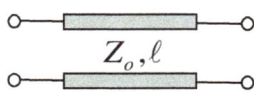

The most proper symbol for a transmission line in high frequency circuits would be one that depicts both top and bottom conductors, as shown in figure.

Two parameters should always be included with the symbol: the characteristic impedance of the transmission line Z_o, and the line length. The characteristic impedance will be defined later in the chapter.

Fig.(8.6) – An equivalent circuit for an infinitesimally short piece of transmission line.

A short piece of ideal lossless transmission line of length Δz can be modeled in the schematic shown in Fig.(8.6). One can visualize this easily with a short piece of coaxial cable in which the electric field is going radially to the ground shield (represented by the shunt capacitor), and the center conductor with magnetic field going around it (represented by the series inductor). The term "shunt" really means going to the ground connection, and is used very often.

The voltage on the right, $V(z+\Delta z)$, is related to the voltage on the right by Kirchhoff's voltage rule:

$$V(z+\Delta z,t) = V(z,t) - (L\Delta z)\frac{\partial i(z,t)}{\partial t} \quad \text{Eqn.(8.1)}$$

From calculus, we learned to expand $V(z+\Delta z)$ in Taylor series and keeping only the first term as $\Delta z \to 0$.

$$V(z+\Delta z,t) \approx V(z,t) + \frac{\partial V(z,t)}{\partial z}\Delta z \quad \text{Eqn.(8.2)}$$

Equate both equations, we get a coupling equation to relate voltage and current.

$$-L\frac{\partial i(z,t)}{\partial t} = \frac{\partial V(z,t)}{\partial z} \quad \text{Eqn.(8.3)}$$

Now, apply the junction rule of current to the schematic in Fig.(8.6) and simplify:

$$i(z+\Delta z,t) - i(z,t) = -(C\Delta z)\frac{\partial V(z,t)}{\partial t} \approx \frac{\partial i(z,t)}{\partial z}\Delta z$$

$$-C\frac{\partial V(z,t)}{\partial t} = \frac{\partial i(z,t)}{\partial z} \quad \text{Eqn.(8.4)}$$

which is another coupling equation. Combining Equations (8.3) and (8.4) yield 2 wave equations. For example, by taking the time derivative of Eqn.(8.3) and substitute Eqn.(8.4) into it, we get:

$$-L\frac{\partial^2 i}{\partial t^2} = \frac{\partial^2 V}{\partial t \partial z} = \frac{\partial}{\partial z}\left(-\frac{1}{C}\frac{\partial i}{\partial z}\right) = -\frac{1}{C}\frac{\partial^2 i}{\partial z^2}$$

$$\frac{\partial^2 i}{\partial z^2} = LC\frac{\partial^2 i}{\partial t^2} \quad \text{Eqn.(8.5)}$$

which is a wave equation traveling in the ±z direction with a velocity = $1/\sqrt{(LC)}$. Again, Appendix D provides a more detailed review on traveling wave and its differential equations.

Likewise, we can take the time derivative of Eqn.(8.4) and substitute Eqn.(8.3) into it. The result is a voltage wave with the same speed:

$$-C\frac{\partial^2 V}{\partial t^2} = \frac{\partial^2 i}{\partial t \partial z} = \frac{\partial}{\partial z}\left(-\frac{1}{L}\frac{\partial V}{\partial z}\right) = -\frac{1}{L}\frac{\partial^2 V}{\partial z^2}$$

$$\frac{\partial^2 V}{\partial z^2} = LC\frac{\partial^2 V}{\partial t^2} \qquad \text{Eqn.(8.6)}$$

The solution of the voltage wave equation (8.6) can be written as:

$$V(z,t) = V_o^+ e^{j(\omega t - \beta z)} + V_o^- e^{j(\omega t + \beta z)} \qquad \text{Eqn.(8.7)}$$

which is a combination of a forward going wave (to the +z direction) and a reverse wave (to the –z direction). The amplitudes V_o^+ and V_o^- are to be determined by the initial conditions and the boundary conditions. Usually, they are interpreted as the incident wave and the reflected wave amplitudes.

Similarly, the current wave can be written as:

$$i(z,t) = I_o^+ e^{j(\omega t - \beta z)} - I_o^- e^{j(\omega t + \beta z)} \qquad \text{Eqn.(8.8)}$$

where I_o^+ and I_o^- are the incident and reflected current amplitudes. Since the amplitudes can be positive or negative, the negative sign in front of I_o^- is just arbitrary and makes no different mathematically. However, there are a couple of reasons to write the negative sign. One of them is due to the fact that the reflected electric field and magnetic field are intrinsically opposite in sign as shown in Fig.(7.1). If the reflected electric field is in-phase with the incident E-field, the reflected magnetic field is 180 degrees out of phase with respect to the incident H-field. Electric field is associated with voltage and magnetic field is with current. So it makes sense to write the voltage wave and current wave differently. The second reason is the convenience of writing the formula in a compact way as illustrated in the next section.

The important lesson here is that by defining a simple equivalent circuit for a piece of transmission line with time-varying field, the wave properties of the voltage and current are obtained. These waves or signals propagate through the transmission line along the guided direction. It is essentially an alternative description of the same electromagnetic waves described in Chapter 6. More of the correlations will be outlined later in the chapter.

8.4 Characteristic Impedance

Voltage and current in a transmission line are traveling waves stated in Equations (8.7) and (8.8). And the voltage and current are related by Equations (8.3) and (8.4). Putting them together, we can relate the voltage and current amplitudes.

$$\frac{\partial i}{\partial z} = -j\beta I_o^+ e^{j(\omega t - \beta z)} - j\beta I_o^- e^{j(\omega t + \beta z)}$$

$$\frac{\partial i}{\partial z} = -C\frac{\partial V}{\partial t} = -C\left[j\omega V_o^+ e^{j(\omega t - \beta z)} + j\omega V_o^- e^{j(\omega t + \beta z)}\right]$$

$$\beta I_o^+ = C\omega V_o^+$$

$$\beta I_o^- = C\omega V_o^-$$

$$V_o^\pm = \frac{\beta}{C\omega} I_o^\pm = \frac{\sqrt{LC}}{C} I_o^\pm = \sqrt{\frac{L}{C}} I_o^\pm \equiv Z_o I_o^\pm \qquad \text{Eqn.(8.9)}$$

The last equation relates the current and voltage amplitude. The ratio of V/I must have the meaning of impedance and unit of ohms. That impedance is called the characteristic impedance of the transmission line.

The choice of the negative sign in the current wave equation Eqn.(8.8) also simplifies the voltage / current ratio to be stated in this simple form. Otherwise, extra signs and conditions would have to be included in Eqn.(8.9).

So what is the characteristic impedance exactly? When we say the coaxial cable is 75 ohms, it does not mean the electrical resistance of the cable is 75Ω. If we connect a 75Ω coaxial cable with another one to make a longer cable, the overall coaxial cable is still 75Ω. Characteristic impedances do not add up. A 75Ω cable plus another 75Ω cable is still a 75Ω cable! We will explain this impedance concept further when we discuss reflection coefficient in a later section.

For all communication systems, the system characteristic impedance is always 50 ohms. All the equipment used in test and measurement are all in one standard impedance. The only other industrial standard is the 75 Ω which is almost used exclusively for commercial televisions. So the coaxial cable from the back of our cable TV set is 75Ω.

8.5 Circuit representation and electromagnetic wave description

The voltage and current wave properties in a transmission line are very similar to the electromagnetic wave properties in a medium. In fact, there is a one-to-one corresponding relation between them. In an earlier chapter, we have the differential equations for the E field and the H field, the explicit wave solutions, the velocity of the wave traveling and the intrinsic impedance of the wave in a medium, as stated below:

$$\nabla^2 \begin{pmatrix} \vec{E} \\ \vec{H} \end{pmatrix} = \mu\varepsilon \frac{\partial^2}{\partial t^2} \begin{pmatrix} \vec{E} \\ \vec{H} \end{pmatrix}$$

$$\vec{E}(z,t) = \vec{E}_{oi} e^{j(\omega t - \vec{k}_i \cdot \vec{r})} + \vec{E}_{or} e^{j(\omega t - \vec{k}_r \cdot \vec{r})}$$

$$\vec{H}(z,t) = \vec{H}_{oi} e^{j(\omega t - \vec{k}_i \cdot \vec{r})} - \vec{H}_{or} e^{j(\omega t - \vec{k}_r \cdot \vec{r})}$$

$$v = \frac{1}{\sqrt{\mu\varepsilon}}$$

$$\eta = \sqrt{\frac{\mu}{\varepsilon}}$$

Eqn.(8.10)

Similarly, here we have the counterparts for the voltage and current waves:

$$\frac{\partial^2}{\partial z^2} \begin{pmatrix} V \\ i \end{pmatrix} = LC \frac{\partial^2}{\partial t^2} \begin{pmatrix} V \\ i \end{pmatrix}$$

$$V(z,t) = V_o^+ e^{j(\omega t - \beta z)} + V_o^- e^{j(\omega t + \beta z)}$$

$$i(z,t) = I_o^+ e^{j(\omega t - \beta z)} - I_o^- e^{j(\omega t + \beta z)}$$

$$v = \frac{1}{\sqrt{LC}}$$

$$Z_o = \sqrt{\frac{L}{C}}$$

Eqn.(8.11)

Their relations are very apparent. The voltage wave is an alternate representation of the electric field oscillation, and the current wave is related to the magnetic field circulation. The inductance per length "L" is introduced to describe the magnetic property of the transmission line. Similarly, the permeability "μ" is the magnetic property of the medium. The capacitance per length "C" describes the electric property of the transmission line, whereas the permittivity "ε" contains the electrical property of the medium. Hence, the wave velocity and impedance take on the same form in both languages. Most engineers are more comfortable working with voltage and current instead of electromagnetic field. With voltage and current, engineers can use many commercial circuit analysis tools to help design and simulate RF circuits, Nonetheless, there are situations (such as in waveguide propagation) where the electromagnetic description in Eqn.(8.10) is more intuitive.

8.6 Reflection at the load

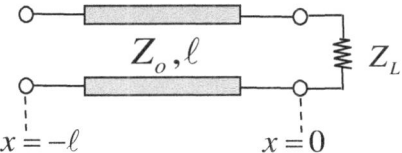

Fig.(8.7) – A load with impedance Z_L is connected to the end of a transmission line.

A load is connected at the end of a transmission line where x = 0. It is a convention that x is positive to the right and negative to the left. Therefore, the other end of the transmission line is at $x = -\ell$. The physical interpretation of any result of course does not depend on how we choose our coordinates.

The voltage and current waves inside the transmission line can be expressed using Eqn.(8.11). Evaluate them at x = 0 would give us the voltage and current relation at the load.

$$V(x) = V_o^+ e^{-j\beta x} + V_o^- e^{j\beta x}$$

$$i(x) = I_o^+ e^{-j\beta x} - I_o^- e^{j\beta x}$$

$$V(0) \equiv V_L = V_o^+ + V_o^-$$

$$i(0) \equiv I_L = I_o^+ - I_o^- = \frac{1}{Z_o}\left(V_o^+ - V_o^-\right)$$

$$\frac{V_L}{I_L} \equiv Z_L = Z_o \left(\frac{V_o^+ + V_o^-}{V_o^+ - V_o^-}\right)$$

$$Z_L\left(V_o^+ - V_o^-\right) = Z_o\left(V_o^+ + V_o^-\right)$$

$$V_o^+\left(Z_L - Z_o\right) = V_o^-\left(Z_L + Z_o\right)$$

$$\frac{V_o^-}{V_o^+} \equiv \Gamma_L = \frac{Z_L - Z_o}{Z_L + Z_o} \qquad \text{Eqn.(8.12)}$$

The load reflection is defined here as the ratio of V_o^- to V_o^+ which indicates the fractional voltage wave amplitude being reflected at the load. Equation (8.12) is similar to Eqn.(7.9) which stated that reflection coefficient is equal to the difference of the impedance divided by the sum. Here, the impedance are the load impedance and the characteristic impedance of the line. Simply put, if there is a difference in impedance, some of the wave or signal will be reflected.

This is what the characteristic impedance of a transmission line stands for. If a transmission line is terminated with a load impedance equal to its characteristic impedance ($Z_L = Z_o$), no signal will be reflected. All the power will be delivered to the load.

The impedance is often normalized to a constant for a number of reasons in future discussion. At this point, it is not so important to do so, but readers should be aware of this convention while reading other references or articles. Let us define a normalized impedance (denoted with a bar on top of the letter Z) as an impedance divided by the characteristic impedance of the transmission line. Again, here we can divide the impedance by any number, but the convention here is to use the Z_o of the line.

$$\bar{Z} \equiv \frac{Z}{Z_o}$$

$$\Gamma_L = \frac{\bar{Z}_L - 1}{\bar{Z}_L + 1}$$
Eqn.(8.13)

The reflection coefficient in Eqn.(8.13) is somewhat less cumbersome than that in Eqn.(8.12). It is easier to deal with a small number than a bigger complex number in general.

Example (8.1) - Suppose the cable connecting the roof top antenna and your TV was gone one day. And you have an extra 50 ohms coaxial cable from the communication lab that you can use. So you modify the cable and solder it to a pair of F connectors (the kind that cable TV uses) and make the connection between the antenna and the TV set. Would that work?

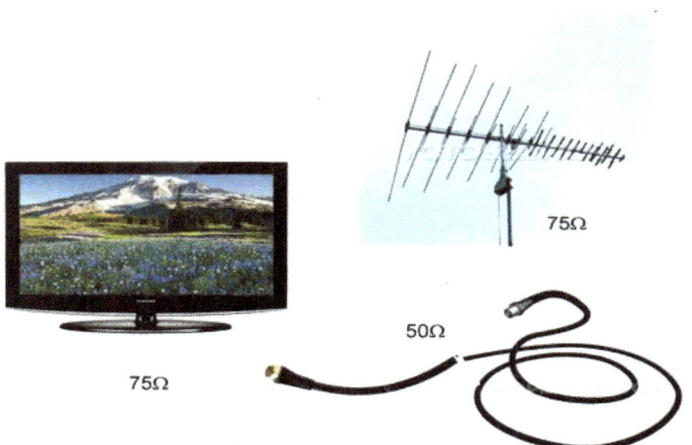

Well, it probably would not work very well.

As the antenna receives a signal and convert it into a 75Ω wave traveling down, it sees a change in impedance to a 50Ω line, Eqn.(8.12) states there must be a reflection (20% amplitude) so part of the signal would not go through.

The rest of the signal travels further through the line and see another discontinuity in impedance, so another 20% of the amplitude goes away again. So, more than 1/3 of the amplitude would not go through. The signal arrives to the TV would be much weaker.

8.7 Input impedance

Impedance is the ratio of voltage over current. With the voltage and current being a wave and not having a constant amplitude, the impedance would appear differently depends on the length of the transmission line among other things.

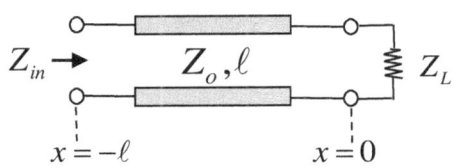

Let us examine Fig.(8.7) again.

What is the impedance looking in from the other side of the transmission line? Starting with the voltage and current wave equations:

$$V(x) = V_o^+ e^{-j\beta x} + V_o^- e^{j\beta x}$$
$$i(x) = I_o^+ e^{-j\beta x} - I_o^- e^{j\beta x}$$

$$Z_{in} \equiv \frac{V_{in}}{I_{in}} = \frac{V(-\ell)}{i(-\ell)} = \frac{V_o^+ e^{j\beta\ell} + V_o^- e^{-j\beta\ell}}{\frac{1}{Z_o}\left(V_o^+ e^{j\beta\ell} - V_o^- e^{-j\beta\ell}\right)}$$

$$Z_{in} = Z_o \left(\frac{e^{j\beta\ell} + \Gamma_L e^{-j\beta\ell}}{e^{j\beta\ell} - \Gamma_L e^{-j\beta\ell}}\right)$$

$$\overline{Z}_{in} = \frac{e^{j\beta\ell} + \Gamma_L e^{-j\beta\ell}}{e^{j\beta\ell} - \Gamma_L e^{-j\beta\ell}} \qquad \text{Eqn.(8.14)}$$

Here, we have used the definition of the normalized impedance, and the load reflected defined in Eqn.(8.12). With a little algebra:

$$\tan x = \frac{\sin x}{\cos x} = \frac{e^{jx} - e^{-jx}}{j(e^{jx} + e^{-jx})}$$
$$(j\tan x)(e^{jx} + e^{-jx}) = e^{jx} - e^{-jx}$$
$$e^{-jx}(1 + j\tan x) = e^{jx}(1 - j\tan x)$$
$$e^{jx} = e^{-jx}\frac{(1 + j\tan x)}{(1 - j\tan x)} \qquad \text{Eqn.(8.15)}$$

Substitute this and Eqn.(8.13) into Eqn.(8.14), we get:

$$\overline{Z}_{in} = \frac{e^{-j\beta\ell}\left(\frac{1 + j\tan\beta\ell}{1 - j\tan\beta\ell}\right) + \left(\frac{\overline{Z}_L - 1}{\overline{Z}_L + 1}\right)e^{-j\beta\ell}}{e^{-j\beta\ell}\left(\frac{1 + j\tan\beta\ell}{1 - j\tan\beta\ell}\right) - \left(\frac{\overline{Z}_L - 1}{\overline{Z}_L + 1}\right)e^{-j\beta\ell}}$$

$$\overline{Z}_{in} = \frac{(\overline{Z}_L + 1)(1 + j\tan\beta\ell) + (\overline{Z}_L - 1)(1 - j\tan\beta\ell)}{(\overline{Z}_L + 1)(1 + j\tan\beta\ell) - (\overline{Z}_L - 1)(1 - j\tan\beta\ell)}$$

$$\bar{Z}_{in} = \frac{2(\bar{Z}_L + j\tan\beta\ell)}{2(1 + j\bar{Z}_L \tan\beta\ell)}$$

$$\bar{Z}_{in} = \frac{\bar{Z}_L + j\tan\beta\ell}{1 + j\bar{Z}_L \tan\beta\ell} \qquad \text{Eqn.(8.16)}$$

$$Z_{in} = Z_o\left(\frac{Z_L + jZ_o \tan\beta\ell}{Z_o + jZ_L \tan\beta\ell}\right) \qquad \text{Eqn.(8.17)}$$

Equations (8.16) and (8.17) are the same except Eqn.(8.16) is normalized (divide every term by Z_o). These two equations are often the starting point of solving most transmission line problem. We will examine the consequence of this equation in the next few sections.

Example (8.2) – In Fig.(8.7), if $Z_o = 50\Omega$, $Z_L = 100\,\Omega$, what is Z_{in}? If the transmission line length is $\lambda/8$? $\lambda/4$? $\lambda/2$? What if $Z_o = Z_L = 50\Omega$? Would the result be different?

Note: Line length is often expressed in terms of lamda or electrical length, because it is always used with tangent function in Equations (8.16) and (8.17).

$$\ell = \frac{\lambda}{8}$$

$$\beta\ell = \frac{2\pi}{\lambda} \cdot \frac{\lambda}{8} = \frac{\pi}{4} = 45°$$

So, $\lambda/8$ is equivalent to $45°$ in electrical length. This, of course, corresponds to the frequency of signal if given.

$$Z_L = 100\Omega$$

$$\bar{Z}_L = \frac{Z_L}{Z_o} = \frac{100}{50} = 2$$

$$\bar{Z}_{in} = \frac{\bar{Z}_L + j\tan\beta\ell}{1 + j\bar{Z}_L \tan\beta\ell} = \frac{2 + j\tan 45°}{1 + j2\tan 45°} = \frac{2+j}{1+j2} = \frac{\sqrt{5}\angle 26.6°}{\sqrt{5}\angle 63.4°} = 1\angle -36.8°$$

$$Z_{in} = Z_o\bar{Z}_{in} = 50\angle -36.8° = 40 - j30\,\Omega$$

For line length = $\lambda/4$? (Electrical length = $90°$) This is called a quarter-wave. Quarter-wave is very special because $\tan(90°)$ is undefined! This special line length is used all the time and will be further discussed in the many later sections. For now, let us resolve the "undefined" issue of this quarter-wave. The limit of Eqn.(8.16) as the line length approach quarter-wave is:

$$\bar{Z}_{in} = \lim_{\beta\ell \to \pi/2}\left(\frac{\bar{Z}_L + j\tan\beta\ell}{1 + j\bar{Z}_L \tan\beta\ell}\right) \approx \frac{j\tan\beta\ell}{j\bar{Z}_L \tan\beta\ell} = \frac{1}{\bar{Z}_L}$$

Eqn.(8.18)

Equation (8.18) gives the input impedance of a load connected to a quarter-wave. We will come back to this equation often for future discussion. For this problem, the input impedance can be explicitly calculated:

$$\bar{Z}_{in} = \frac{1}{\bar{Z}_L} = \frac{1}{2}$$

$$Z_{in} = 50\left(\frac{1}{2}\right) = 25\Omega$$

For line length = $\lambda/2$, electrical length is 180°, tan(180°) = 1, so $Z_{in} = Z_L = 100\Omega$.

The lesson of this example is that: although the load is a simple 100Ω resistor, the input impedance on the other side of the transmission line can appear to be very different, and varied depends on the length of the transmission line. This is very different than low frequency circuits where line length would not normally change the input impedance.

What if $Z_o = Z_L = 50\Omega$?

$$Z_L = 50\Omega$$

$$\bar{Z}_L = \frac{Z_L}{Z_o} = \frac{50}{50} = 1$$

$$\bar{Z}_{in} = \frac{\bar{Z}_L + j\tan\beta\ell}{1 + j\bar{Z}_L \tan\beta\ell} = \frac{1 + j\tan\beta\ell}{1 + j\tan\beta\ell} = 1$$

$$Z_{in} = Z_o = 50\Omega$$

In other words, if the load impedance is matched with the characteristic impedance of the transmission line, the line length has no effect on the input impedance regardless the length of the transmission line. This is a desired property. And it is the ultimate goal for RF engineers to design their components and systems to 50 ohms so that the interconnect transmission line length does not degrade the performance of the system.

Low frequency circuit designs never have to be concern about the length of the connecting wire. This can be explained easier with our transmission line equation (8.16). Low frequency means long wavelength. So, $f \to 0$, $\lambda \to \infty$, $\beta \to 0$.

$$\bar{Z}_{in} = \lim_{f \to 0} \frac{\bar{Z}_L + j\tan\beta\ell}{1 + j\bar{Z}_L \tan\beta\ell} \approx \frac{\bar{Z}_L}{1} = \bar{Z}_L$$

$$Z_{in} = Z_L$$

So the line length and line impedance do not matter at all at low frequency circuits.

8.8 Input reflection

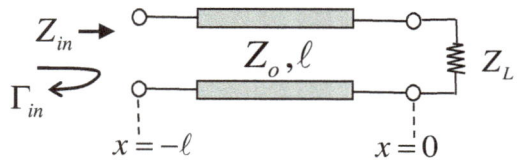

Since the input impedance can change significantly depending on the line length and the characteristic impedance of the line, so would the reflection at the input end of the line. Reflection coefficient is the difference of the impedances divide by their sum.

$$\Gamma_{in} = \frac{Z_{in} - Z_{left}}{Z_{in} + Z_{left}} \qquad \text{Eqn.(8.19)}$$

where Z_{left} is the impedance to the left of the transmission line in the figure. Usually this would be connected to another 50 Ω cable from the equipment so Z_{left} would be just 50 Ω. However, if it is connected to some other devices, Z_{left} would be the characteristic impedance of that transmission line, and not the equivalent impedance of the device.

<u>Example (8.3)</u> – In Fig.(8.7), if $Z_o = 50Ω$, $Z_L = 100$ Ω, and the transmission line length is $\lambda/8$, what is Γ_L and Γ_{in} if this is a 50Ω communication system? Would the result be different if this is a 75Ω, TV system?

$$\bar{Z}_L = \frac{Z_L}{Z_o} = \frac{100}{50} = 2$$

$$\bar{Z}_{in} = \frac{\bar{Z}_L + j\tan\beta\ell}{1 + j\bar{Z}_L \tan\beta\ell} = \frac{2 + j\tan 45°}{1 + j2\tan 45°} = \frac{2+j}{1+j2} = \frac{\sqrt{5}\angle 26.6°}{\sqrt{5}\angle 63.4°} = 1\angle -36.8°$$

$$Z_{in} = Z_o \bar{Z}_{in} = 50\angle -36.8° = 40 - j30 \, \Omega$$

$$\Gamma_L = \frac{Z_L - Z_o}{Z_L + Z_o} = \frac{100-50}{100+50} = \frac{1}{3}$$

$$\Gamma_{in} = \frac{Z_{in} - 50}{Z_{in} + 50} = \frac{40 - j30 - 50}{40 - j30 + 50} = \frac{31.623\angle -108°}{94.868\angle -18°} = \frac{1}{3}\angle -90° = -\frac{j}{3}$$

Note: In a 50 Ω system with a 50 Ω line, the load reflection and input reflection have the same magnitude. i.e., the input reflection is solely due to the reflection at the load. The 90 degree phase shift is due to the extra length the reflection has to travel back and forth through the 45-degrees transmission line.

If this is a 75 Ω system, then the input end would pick up extra reflection:

$$\Gamma_{in} = \frac{Z_{in} - 75}{Z_{in} + 75} = \frac{40 - j30 - 75}{40 - j30 + 75} = \frac{46.1\angle -139.4°}{118.8\angle -14.6°} = 0.388\angle -125°$$

$$\Gamma_{in} > \Gamma_L$$

8.9 Voltage Standing Wave Ratio (VSWR)

In Section 7.4, we defined SWR when the electric field incident onto a boundary and partially reflected. The incident wave and the reflected wave form a standing wave in front of the boundary. Here the voltage wave is partially reflected whenever we have a change in impedance. So a voltage standing wave is formed in the transmission line. The ratio of the voltage maximum to voltage minimum is called the VSWR.

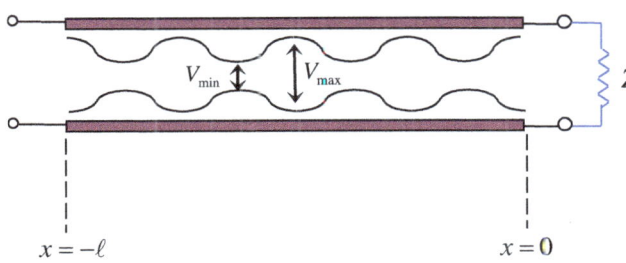

Fig.(8.8) – A voltage standing wave is formed in a transmission due to the partial reflection at the load.

$$V(x) = V_o^+ e^{-j\beta x} + V_o^- e^{j\beta x}$$

$$V(x) = V_o^+ e^{-j\beta x}\left(1 + \Gamma_L e^{2j\beta x}\right)$$

$$|V| = |V_o^+|\left|1 + \Gamma_L e^{2j\beta x}\right|$$

$$\Gamma_L \equiv \rho e^{j\theta} \quad \text{(Define the complex load reflection in polar form.)}$$

$$|V| = V_o^+ \left|1 + \rho e^{j(\theta + 2\beta x)}\right|$$

$$|V| = V_o^+ \sqrt{(1 + \rho\cos(\theta + 2\beta x))^2 + \rho^2 \sin^2(\theta + 2\beta x)}$$

$$|V| = V_o^+ \sqrt{1 + 2\rho\cos(\theta + 2\beta x) + \rho^2}$$

$$|V| = V_o^+ \sqrt{(1+\rho)^2 - 2\rho(1 - \cos(\theta + 2\beta x))}$$

$$|V| = V_o^+ \sqrt{(1+\rho)^2 - 4\rho\sin^2\left(\frac{\theta + 2\beta x}{2}\right)}$$

This is the magnitude of voltage, as a function of position, in the transmission line. The maxima and minima of the voltage can be identified by setting the sine term to 1 or 0, respectively.

$$V_{max} = V_o^+ \sqrt{(1+\rho)^2} = V_o^+ (1+\rho)$$

at $\quad \dfrac{\theta + 2\beta x}{2} = \pm n\pi$

$$x_{max} = -[\theta \mp 2n\pi]\frac{\lambda}{4\pi} \qquad \text{Eqn.(8.20)}$$

And the voltage minimum is:

$$V_{min} = V_o^+ \sqrt{(1+\rho)^2 - 4\rho} = V_o^+(1-\rho)$$

at $\quad \dfrac{\theta + 2\beta x}{2} = \pm \dfrac{(2n+1)\pi}{2}$

$$x_{min} = -[\theta \mp (2n+1)\pi]\dfrac{\lambda}{4\pi} \qquad \text{Eqn.(8.21)}$$

Equations (8.20) and (8.21) give the location for all voltage maxima and minima. Note that x must be a negative number according to the convention that x = 0 at the load. The integer n can be any integer provide that $-\ell < x < 0$.

The voltage standing wave ratio (VSWR) can be written in terms of the magnitude of the reflection coefficient ρ:

$$VSWR \equiv \dfrac{V_{max}}{V_{min}} = \dfrac{V_o^+(1+\rho)}{V_o^+(1-\rho)} = \dfrac{1+\rho}{1-\rho} = \dfrac{1+|\Gamma|}{1-|\Gamma|} \qquad \text{Eqn.(8.22)}$$

Note: The standing wave is a pattern setup in the transmission line. So the reflection coefficient is always the reflection in front of (or to the right of) the transmission line. If this is the input VSWR, then the standing wave is assumed to be on the transmission line connecting the device (circuit) to the equipment (not shown in the schematic), and the reflection coefficient is the Γ_{in}.

Input VSWR is one of the key specifications people still use today in the industry. It basically provides the information of the input reflection, which in turns indicates how well does the input impedance match up with the system impedance (50Ω).

VSWR ranges from 1 to infinity. If the impedance matches up with Z_o, there will be no reflection at the connection, $\Gamma = 0$, and VSWR = 1. In case of a open or short circuit, there will be a total reflection, $|\Gamma| = 1$, and VSWR $\rightarrow \infty$. A desirable VSWR is very close to one, i.e., perfectly matched circuit. In practice, input VSWR in commercial hardware always range from 1.1 to 2.

With a little algebra, Eqn.(8.22) can be written as:

$$\rho = |\Gamma| = \dfrac{VSWR - 1}{VSWR + 1} \qquad \text{Eqn.(8.23)}$$

Recall that the reflection coefficient is:

$$\Gamma = \dfrac{Z - Z_o}{Z + Z_o} = \dfrac{\bar{Z} - 1}{\bar{Z} + 1} \qquad \text{Eqn.(8.24)}$$

It seems to suggest that the VSWR also takes on the meaning of the normalized impedance, but only if the impedance is a pure real number. In general this is NOT the case. Nevertheless, it gives a rough idea of what the system can tolerate. For example, VSWR of 1.1 is consider a good match, which roughly corresponds to a normalized impedance of 1.1, or 55 Ω. A bad VSWR would be more than 2, which translate to an impedance of more than 100 Ω. Again, this is NOT correct to assume the impedance is real. There is often a reactance part to it that we cannot ignore.

8.10 Return Loss

Return Loss is defined as the logarithm scale of this magnitude ρ:

$$RL = -20\log_{10}|\Gamma| = -20\log\rho \qquad \text{Eqn.(8.24)}$$

Return loss, reflection coefficient, and VSWR are closely related and basically convey the same information. They all indicate how well does the input impedance match to the system impedance. Return loss, differs from the other two, applies only to the terminals. Usually they are referred to as input or output return loss. Input return loss has becoming more popular than the input VSWR, primarily due to its similarity to the other parameters used in the system language (such as gain, insertion loss, or noise figure) that are all in the same logarithmic scale.

Below is a table showing the relation between VSWR, ρ, and return loss for a few common values:

VSWR	$\rho = |\Gamma|$	Return Loss	
1.0	0	∞	Perfectly matched impedance
1.1	0.0476	26 dB	Good impedance matching
1.5	0.2	14 dB	Average impedance matching
2.0	0.33	9.5 dB	Poor impedance matching
∞	1	0 dB	Open or short circuit

<u>Example (8.4)</u> - In the figure below, the transmission line impedance is $Z_o = 25\Omega$ and line length is 1.4λ. The load impedance $Z_L = 75 - j50\ \Omega$. Calculate the VSWR in the 25Ω line, the input VSWR, and the input return loss, knowing this is a 50Ω communication system. Also locate all the voltage maxima in the line.

$$\bar{Z}_L = \frac{Z_L}{Z_o} = \frac{75-j50}{25} = 3-j2$$

$$\Gamma_L = \frac{\bar{Z}_L-1}{\bar{Z}_L+1} = \frac{2-j2}{4+j2} = 0.63\angle-18°$$

$$VSWR_{LINE} = \frac{1+\rho}{1-\rho} = \frac{1.63}{0.37} = 4.4$$

The voltage maxima are given by Eqn.(8.20), with θ is the phase in Γ_L = -18° = -0.322 (radian). So the first negative x_{max} is when n = +1, and 0.5λ thereafter.

$$x_{max \atop first} = -[\theta \mp 2n\pi]\frac{\lambda}{4\pi} = -[-0.322+2\pi]\frac{\lambda}{4\pi} = -0.474\lambda$$

There are 2 maxima totally within the physical length of 1.4λ. They are located to the left of the load at -0.474λ, and -0.974λ.

$$\tan\beta\ell = \tan\left[\frac{2\pi}{\lambda}(1.4\lambda)\right] = \tan\left[\frac{2\pi}{\lambda}(0.4\lambda)\right] = \tan(0.8\pi) = \tan(144°) = -0.727$$

$$\bar{Z}_{in} = \frac{\bar{Z}_L+j\tan\beta\ell}{1+j\bar{Z}_L\tan\beta\ell} = \frac{3-j2+j(-0.727)}{1+j(3-j2)(-0.727)} = 0.925+j1.57$$

$$Z_{in} = Z_o\bar{Z}_{in} = (25)\bar{Z}_{in} = 23.1-j39.2\,\Omega$$

$$\Gamma_{in} = \frac{Z_{in}-50}{Z_{in}+50} = \frac{23.1-j39.2-50}{23.1-j39.2+50} = 0.573\angle 96.2°$$

$$RL = -20\log|\Gamma_{in}| = 4.8 dB$$

$$VSWR_{INPUT} = \frac{1+\rho}{1-\rho} = \frac{1+0.573}{1-0.573} = 3.7$$

8.11 Stubs

A stub is a piece of transmission line that does not appear to be connected on one or both ends. Actually it is always connected to the signal path at one end either thru direct contact or capacitive coupling (i.e. no physical contact), while leaving the other end open or shorted. They are called open stubs or short stubs, represented by the following symbols:

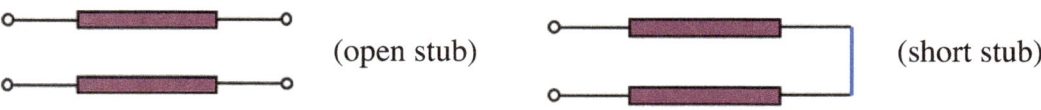

(open stub)   (short stub)

And they are called series stubs if they are connected in series.

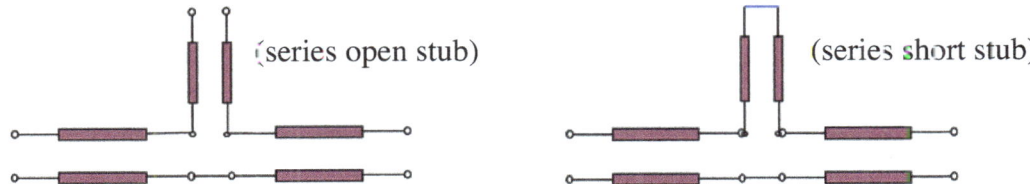

Similarly, there are shunt open stubs and shunt short stubs.

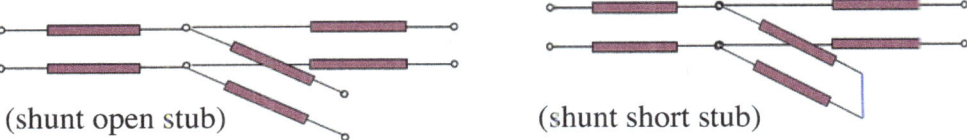

At microwave and millimeter wave frequencies, chip capacitors and inductors do not work well due to their inherent parasitic and self-resonant. Instead, circuit designers used stubs to replace capacitors and inductors. Below are a few examples.

Fig.(8.9) – A edge-coupled filter on microstrip. Input and output ports are on the 2 ends. The 4 main lines are open stubs. The middle 2 stubs are not physically connected to anything.

Fig.(8.10) – A interdigital microstrip filter on a carrier. Input and output ports are on the 2 ends. There are 8 pairs of stubs shorted to the carrier. They are the shunt short stubs. At the designed frequency, it provides a perfect transmission path straight through.

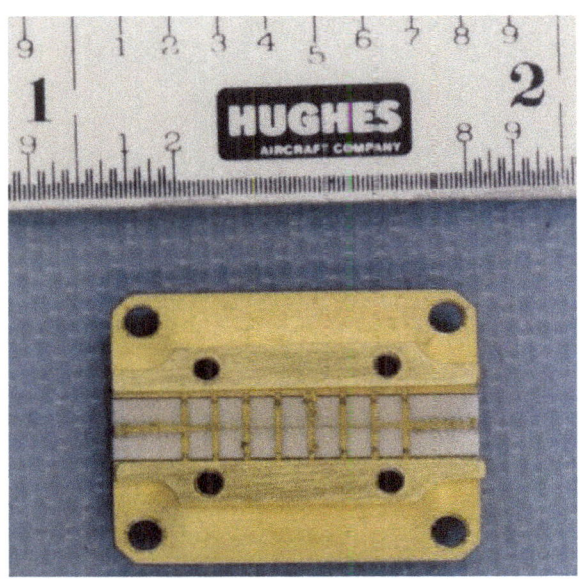

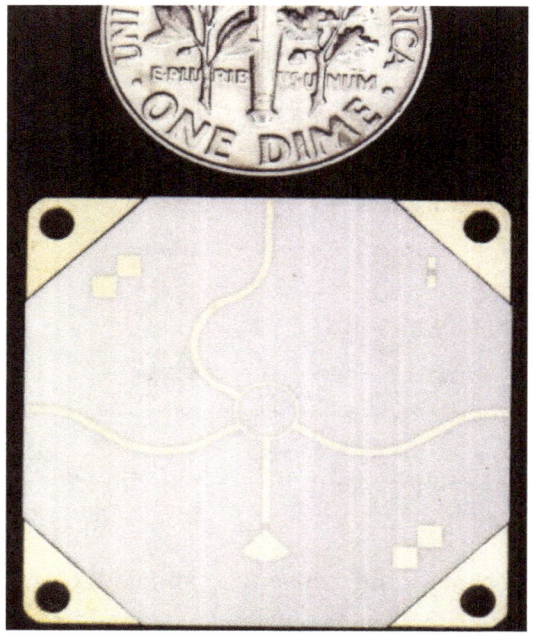

Fig.(8.11) – A "rat race" is a 3-port network that provide coupling or power combining functions. In the lower portion of the circuit, the spade-like is a shunt open stub in the radial form, The little dark square right above the stub is an etched resistor. The open stub is a quarter-wave stub. Its function is to provide a short connection to the resistor even though no part of the circuit is physically connected to the ground! This is one of the fascinating topics covered in Microwave Engineering.

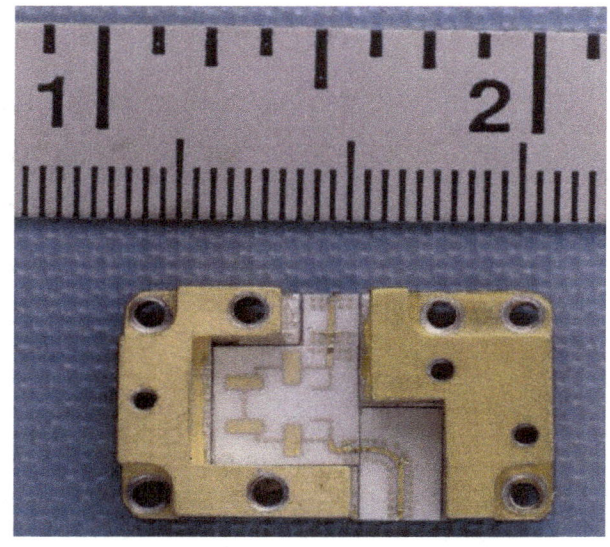

Fig.(8.12) – A low pass filter on microstrip. Similar to the other photographs shown, there is no chip inductor or capacitor in the design. All inductance and capacitance are realized by transmission lines ONLY. The tuning stubs close to the input and output ports are shunt open stubs.

8.11.1 Short stub

A short stub is a piece of transmission line with zero load impedance. The input impedance of a short stub can be found by setting $Z_L = 0$ in the transmission line equation.

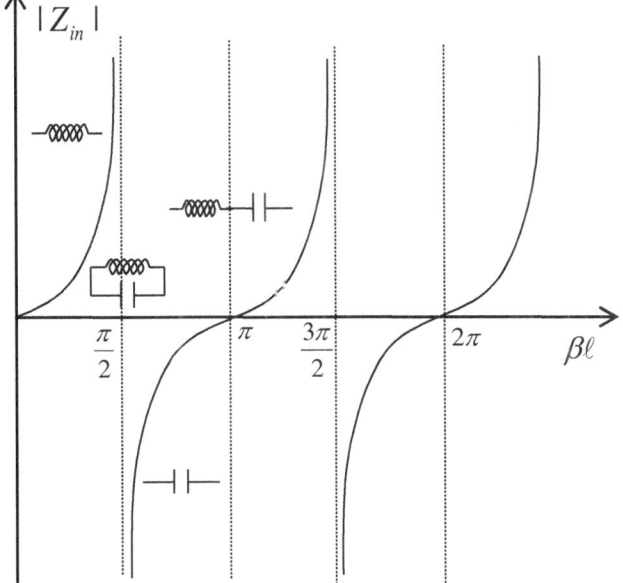

$$Z_L = 0$$

$$\overline{Z}_{in} = \frac{\overline{Z}_L + j\tan\beta\ell}{1 + j\overline{Z}_L \tan\beta\ell}$$

$$\overline{Z}_{sh} = j\tan\beta\ell$$

$$\overline{Y}_{sh} \equiv \frac{1}{\overline{Z}_{sh}} = -j\cot\beta\ell \qquad \text{Eqn.(8.25)}$$

Fig.(8.13) – Input impedance of a short stub.

For a short piece of short stub, the input impedance is a positive pure imaginary number. In other words, the short stub behaves like an inductor: $Z_{coil} = j\omega L$.

So, for a given frequency, we can choose a short stub with the correct length and characteristic impedance to get the inductance that we need.

If the stub length is longer, say between $\pi/2$ to π in electrical length, the input impedance is a pure negative imaginary number. That means the short stub can behave like a capacitor as well; $Z_{cap} = 1/j\omega C$. This is also true for a fix piece of stub, as frequency changes, i.e., β changes, its behavior can change drastically from inductive to capacitive.

For $\beta\ell \to \pi$, $Z_{in} = 0$. This is equivalent to having a series resonator which has the impedance of:

$$Z_{sres} = j\omega L\left(1 - \frac{1}{\omega^2 LC}\right) \qquad \text{Eqn.(8.26)}$$

Lastly, as $\beta\ell \to \pi/2$, Z_{in} is undefined, which corresponding to a tank resonator. The filter shown in Fig.(8.10) is an example of utilizing quarter-wave short stubs. The impedance of each stub is:

$$Z_{pres} = \frac{1}{j\omega C\left(1 - \frac{1}{\omega^2 LC}\right)} \qquad \text{Eqn.(8.27)}$$

The resonator equivalence is particularly useful for filter, matching network, and oscillator designs. Other stub lengths are useful for replacing capacitors and inductors. Since tangent has a period of π, one should expect the circuit behaves similar at higher harmonics. (Remember β is proportional to f.)

8.11.2 Open stub

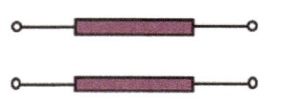

An open stub is a piece of transmission line can be modeled by setting $Z_L \to \infty$ in the transmission line equation.

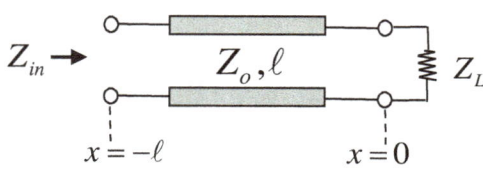

$$Z_L \to \infty$$

$$\bar{Z}_{in} = \lim_{Z_L \to \infty} \frac{\bar{Z}_L + j\tan\beta\ell}{1 + j\bar{Z}_L \tan\beta\ell}$$

$$\bar{Z}_{op} = -j\cot\beta\ell$$

$$\bar{Y}_{op} = j\tan\beta\ell \qquad\qquad \text{Eqn.(8.28)}$$

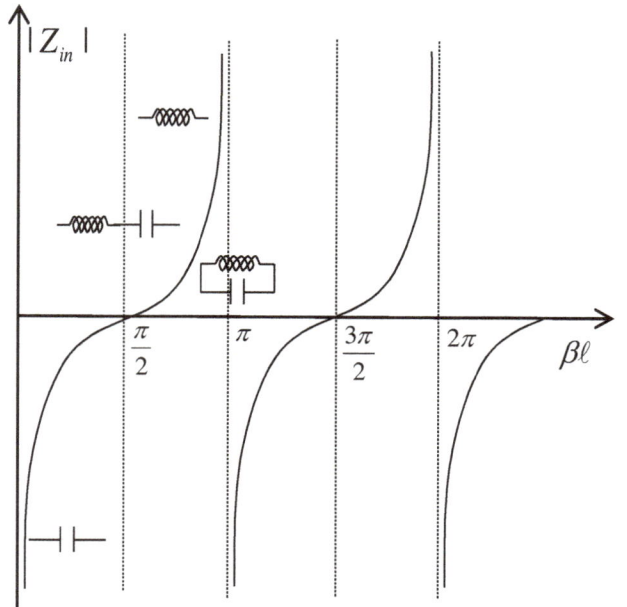

Similarly, a small piece of open stub behaves like a capacitor. But at other line length, it can be a inductor, or a resonator as well.

At quarter-wave (the electrical length is π/2), an open stub is used as a series resonator. A longer stub at half-wave (where its equivalent electrical length is π) behaves as a tank resonator. The corresponding features of an open stub as a function of length (or frequency) is listed in Fig.(8.14).

Fig.(8.14) – Input impedance of a open stub.

Example (8.5) - For the circuit given, find Z_{in} and Γ_{in}.

$$Z_{short} = jZ_o \tan \beta\ell = j(75)\tan(45°) = j75\,\Omega$$

$$\ell = 0.1\lambda$$
$$\beta\ell = (0.1)(2\pi) = 0.2\pi = (0.1)(360°) = 36°$$
$$Z_{open} = -jZ_o \cot \beta\ell = -j(100)\cot(36°) = -j138\,\Omega$$

$$Y_1 = \frac{1}{j75\,\Omega} = -j0.0133$$

$$Y_2 = \frac{1}{-j138\,\Omega} = j0.0073$$

$$Y_{total} = Y_1 + Y_2 = \frac{1}{-j138\,\Omega} = -j0.0060$$

$$Z_{total} = \frac{1}{Y_{total}} = \frac{1}{-j0.006} = j167\,\Omega$$

$$\overline{Z}_{in} = \frac{\overline{Z}_L + j\tan\beta\ell}{1 + j\overline{Z}_L \tan\beta\ell} = \frac{\overline{Z}_L + j\tan\pi}{1 + j\overline{Z}_L \tan\pi} = \overline{Z}_L$$

$$Z_{in} = Z_{total} = j167\,\Omega$$

Unless it is specified, the system impedance is always 50Ω.

$$\Gamma_{in} = \frac{Z_{in} - Z_o}{Z_{in} + Z_o} = \frac{j165 - 50}{j165 + 50} = 1\angle(107° - 73°) = 1\angle 34°$$

8.12 Input admittance

Sometimes it is easier to work with admittance than impedance especially when we are working with shunt elements, or adding parallel circuit elements. For a piece of transmission line, we recall the input impedance from Eqn.(8.15):

$$\overline{Z}_{in} = \frac{\overline{Z}_L + j\tan\beta\ell}{1 + j\overline{Z}_L \tan\beta\ell}$$

$$\overline{Y}_{in} \equiv \frac{1}{\overline{Z}_{in}} = \frac{1 + j\overline{Z}_L \tan\beta\ell}{\overline{Z}_L + j\tan\beta\ell}$$

$$\overline{Y}_{in} = \frac{1 + j(1/\overline{Y}_L)\tan\beta\ell}{1/\overline{Y}_L + j\tan\beta\ell}$$

which can be expressed as:

$$\overline{Y}_{in} = \frac{\overline{Y}_L + j\tan\beta\ell}{1 + j\overline{Y}_L \tan\beta\ell} \qquad \text{Eqn.(8.29)}$$

$$Y_{in} = Y_o\left(\frac{Y_L + jY_o \tan\beta\ell}{Y_o + jY_L \tan\beta\ell}\right) \qquad \text{Eqn.(8.30)}$$

The admittance equations (8.29) and (8.30) are essentially the same as the impedance equations (8.15) and (8.16).

The input reflection coefficient is related to the input admittance directly:

$$\Gamma_{in} = \frac{Z_{in} - Z_1}{Z_{in} + Z_1}$$

$$\Gamma_{in} = \frac{1/Y_{in} - 1/Y_1}{1/Y_{in} + 1/Y_1}$$

$$\Gamma_{in} = \frac{Y_1 - Y_{in}}{Y_1 + Y_{in}} = \frac{1 - \overline{Y}_{in}}{1 + \overline{Y}_{in}} \qquad \text{Eqn.(8.31)}$$

<u>Example (8.6)</u> - Calculate Z_{in}, Γ_{in} and the input return loss of the circuit below. Frequency of operation is 1.592 GHz. System impedance is 50 Ω. What is the VSWR in the 100 Ω transmission line? Locate all the voltage maxima in the transmission line.

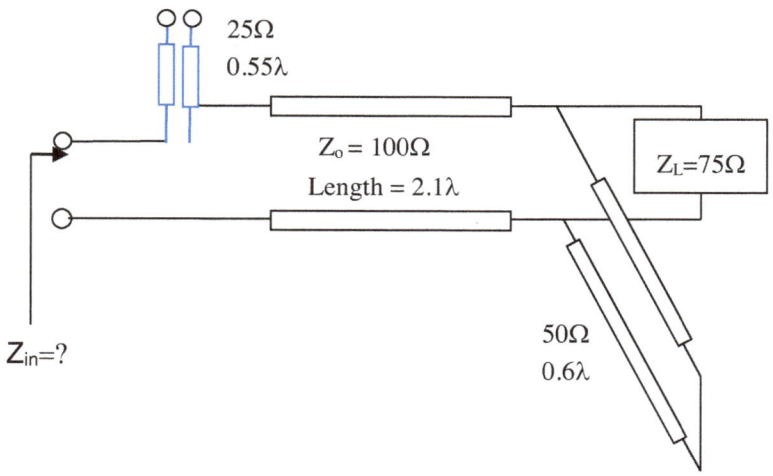

118

$$Y_L = \frac{1}{Z_L} = \frac{1}{75} = 0.013$$

$$Y_{short} = -jY_o \cot(\beta\ell) = -j\left(\frac{1}{50}\right)\cot[(0.6)(360°)] = -j0.028$$

$$Y_1 = Y_{short} + Y_L = 0.013 - j0.028$$

$$Y_{LINE} = \frac{1}{100} = 0.01$$

$$Y_2 = Y_o\left(\frac{Y_L + jY_o \tan\beta\ell}{Y_o + jY_L \tan\beta\ell}\right) = (0.01)\left[\frac{(0.013 - j0.028) + j(0.01)\tan(36°)}{0.01 + j(0.013 - j0.028)\tan(36°)}\right]$$

$$Y_2 = (2.05 - j7.42)\cdot 10^{-3}$$

$$Z_2 = \frac{1}{Y_2} = 34.6 + j125$$

$$Z_{open} = -jZ_o\cot(\beta\ell) = -j(25)\cot[(0.55)(360°)] = -j76.9$$

$$Z_{in} = Z_2 + Z_{open} = (34.6 + j125) - j76.9 = 34.6 + j48.3\,\Omega$$

$$\Gamma_{in} = \frac{Z_{in} - 50}{Z_{in} + 50} = \frac{34.6 + j48.3 - 50}{34.6 + j48.3 + 50} = 0.52\angle 78°$$

$$RL = -20\log\rho = -20\log(0.52) = 5.7\,dB$$

To find VSWR on the 100W transmission line, we have to find the reflection coefficient right after the shunt short stub, Γ_1:

$$\Gamma_1 = \frac{Y_{LINE} - Y_1}{Y_{LINE} + Y_1} = \frac{0.01 - (0.013 - j0.029)}{0.01 + (0.013 - j0.029)} = 0.77\angle 147° = 0.77\angle 2.56(rad)$$

$$VSWR_{LINE} = \frac{1+\rho}{1-\rho} = \frac{1+0.77}{1-0.77} = 7.64$$

The voltage maxima are located at:

$$x_{max} = -[\theta \mp 2n\pi]\frac{\lambda}{4\pi} = -[2.56 \mp 2n\pi]\frac{\lambda}{4\pi}$$

Since θ is > 0, the first maximum would be found at n = 0, and every 0.5λ thereafter.

$$x_{max} = -0.204\lambda, -0.704\lambda, -1.204\lambda, -1.704\lambda$$

Total line length is 2.1λ, so only 4 maxima are found.

Example (8.7) – Power consideration. Let us revisit the circuit in Example (8.3). How much power is delivered to the load? How much power is being reflected?

From Example (8.3), we already calculated:
$$\Gamma_L = \frac{1}{3} \quad \text{and} \quad \Gamma_{in} = \frac{1}{3}\angle -90°.$$

Power reflected at the input port is:
$$P_{\substack{reflected \\ input}} = |\Gamma_{in}|^2 = \left(\frac{1}{3}\right)^2 = \frac{1}{9} = 11\%$$

Power delivered = 89%, or $\quad P_{\substack{thru \\ input}} = 1 - |\Gamma_{in}|^2 = \frac{8}{9} = 89\%$

Note: Γ_L and Γ_{in} have the same magnitude. They refer to the same reflection, so do not double count the reflected power. The power goes through the input port delivers into the load. This is true as long as the transmission line is lossless. In terms of dB, the reflected power is given by the return loss:

$$RL = -20\log|\Gamma_{in}| = -20\log\left(\frac{1}{3}\right) = -9.5 dB$$

The fractional power transmitted is:

$$P_{thru} = 10\log\left[1 - |\Gamma_{in}|^2\right] = 10\log\left(\frac{8}{9}\right) = -1 dB$$

This is also referred to as insertion loss (except the word "loss" absorbed the "-"ve sign.)

$$IL = -10\log\left[1 - |\Gamma_{in}|^2\right] = 10\log\left(\frac{8}{9}\right) = 1 dB$$

which means the power delivered to the load is 1 dB below the input power.

Appendix A: Transformation of selected coordinates

The three most common orthogonal curvilinear coordinate systems we used are the Cartesian coordinates (x, y, z), the Cylindrical coordinates (r, ϕ, z), Spherical coordinates (R, θ, ϕ). Sometimes it is useful to convert one set of coordinates to another. Most elementary calculus textbooks provide the details of the transformations. As always, the best way to write these transformations is by drawing out the coordinates in the same diagram (as shown). Below is a summary of such transformations for quick references.

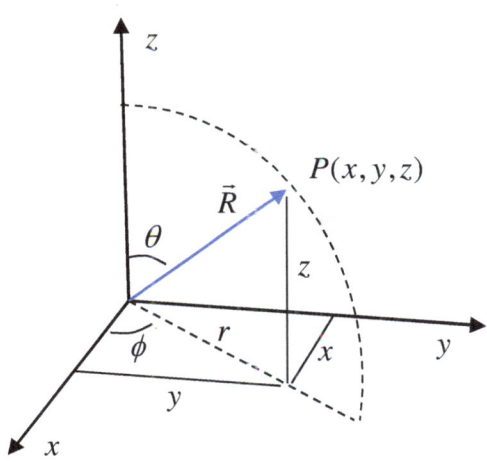

Fig. A – Relationships between coordinates and variables.

	Cartesian Coordinates (x, y, z)	Cylindrical Coordinates (r, ϕ, z)	Spherical Coordinates (R, θ, ϕ)
Cartesian Coordinates (x, y, z)		$x = r\cos\phi$ $y = r\sin\phi$ $z = z$	$x = R\sin\theta\cos\phi$ $y = R\sin\theta\sin\phi$ $z = R\cos\theta$
Cylindrical Coordinates (r, ϕ, z)	$r = \sqrt{x^2 + y^2}$ $\tan\phi = \dfrac{y}{x}$ $z = z$		$r = R\sin\theta$ $\phi = \phi$ $z = R\cos\theta$
Spherical Coordinates (R, θ, ϕ)	$R = \sqrt{x^2 + y^2 + z^2}$ $\tan\theta = \dfrac{\sqrt{x^2 + y^2}}{z}$ $\tan\phi = \dfrac{y}{x}$	$R = \sqrt{r^2 + z^2}$ $\tan\theta = \dfrac{r}{z}$ $\phi = \phi$	

The unit vectors are also related according to the diagram.

	Cartesian Coordinates (x, y, z)	Cylindrical Coordinates (r, φ, z)	Spherical Coordinates (R, θ, φ)
Cartesian Coordinates (x, y, z)		$\hat{x} = \hat{r}\cos\phi - \hat{\phi}\sin\phi$ $\hat{y} = \hat{r}\sin\phi + \hat{\phi}\cos\phi$ $\hat{z} = \hat{z}$	$\hat{x} = \hat{R}\sin\theta\cos\phi + \hat{\theta}\sin\theta\sin\phi - \hat{\phi}\sin\phi$ $\hat{y} = \hat{R}\sin\theta\sin\phi + \hat{\theta}\cos\theta\sin\phi + \hat{\phi}\cos\phi$ $\hat{z} = \hat{R}\cos\theta - \hat{\theta}\sin\theta$
Cylindrical Coordinates (r, φ, z)	$\hat{r} = \hat{x}\cos\phi + \hat{y}\sin\phi$ $\hat{\phi} = -\hat{x}\sin\phi + \hat{y}\cos\phi$ $\hat{z} = \hat{z}$		$\hat{r} = \hat{R}\sin\theta + \hat{\theta}\cos\theta$ $\hat{\phi} = \hat{\phi}$ $\hat{z} = \hat{R}\cos\theta - \hat{\theta}\sin\theta$
Spherical Coordinates (R, θ, φ)	$\hat{R} = \hat{x}\sin\theta\cos\phi + \hat{y}\sin\theta\sin\phi + \hat{z}\cos\theta$ $\hat{\theta} = \hat{x}\cos\theta\cos\phi + \hat{y}\cos\theta\sin\phi - \hat{z}\sin\theta$ $\hat{\phi} = -\hat{x}\sin\phi + \hat{y}\cos\phi$	$\hat{R} = \hat{r}\sin\theta + \hat{z}\cos\theta$ $\hat{\theta} = \hat{r}\cos\theta - \hat{z}\sin\theta$ $\hat{\phi} = \hat{\phi}$	

Appendix B: Differential operations in selected coordinates

In Cartesian coordinates, distance between two nearby points can be written as:

$ds^2 = dx^2 + dy^2 + dz^2.$

In other orthogonal curvilinear coordinates (x_1, x_2, x_3), the same distance can be expressed as:

$$ds^2 = \sum_{i=1}^{3} (h_i dx_i)^2.$$

Here h_i are the scaling factors of the coordinates relative to the Cartesian coordinates, such that the differential distance is: $ds_i = h_i\, dx_i$, and the differential area is: $da_{ij} = (h_i\, h_j)(dx_i\, dx_j)$, and the differential volume is: $dV = (h_1\, h_2\, h_3)(dx_1\, dx_2\, dx_3)$.

Consequently, the generalized differential operators are:

Gradient:
$$\nabla f = \sum_{i=1}^{3} \hat{x}_i \frac{\partial f}{h_i \partial x_i} = \hat{x}_1 \frac{\partial f}{h_1 \partial x_1} + \hat{x}_2 \frac{\partial f}{h_2 \partial x_2} + \hat{x}_3 \frac{\partial f}{h_3 \partial x_3}$$

Divergence:
$$\nabla \cdot \vec{A} = \frac{1}{h_1 h_2 h_3} \left[\frac{\partial}{\partial x_1}(A_1 h_2 h_3) + \frac{\partial}{\partial x_2}(A_2 h_3 h_1) + \frac{\partial}{\partial x_3}(A_3 h_1 h_2) \right]$$

Curl:
$$\nabla \times \vec{A} = \frac{1}{h_1 h_2 h_3} \begin{vmatrix} h_1 \hat{x}_1 & h_2 \hat{x}_2 & h_3 \hat{x}_3 \\ \frac{\partial}{\partial x_1} & \frac{\partial}{\partial x_2} & \frac{\partial}{\partial x_3} \\ h_1 A_1 & h_2 A_2 & h_3 A_3 \end{vmatrix}$$

Laplacian:
$$\nabla^2 = \nabla \cdot \nabla = \frac{1}{h_1 h_2 h_3} \left[\frac{\partial}{\partial x_1}\left(\frac{h_2 h_3}{h_1} \frac{\partial}{\partial x_1}\right) + \frac{\partial}{\partial x_2}\left(\frac{h_3 h_1}{h_2} \frac{\partial}{\partial x_2}\right) + \frac{\partial}{\partial x_3}\left(\frac{h_1 h_2}{h_3} \frac{\partial}{\partial x_3}\right) \right]$$

(Note: the Laplacian can operate on either a scalar function or a vector function.)

Cartesian Coordinates (x, y, z)

$h_1 = h_x = 1$,

$h_2 = h_y = 1$,

$h_3 = h_z = 1$,

$dV = dxdydz$

$$\nabla f = \hat{x}\frac{\partial f}{\partial x} + \hat{y}\frac{\partial f}{\partial y} + \hat{z}\frac{\partial f}{\partial z}$$

$$\nabla \cdot \vec{A} = \frac{\partial A_x}{\partial x} + \frac{\partial A_y}{\partial y} + \frac{\partial A_z}{\partial z}$$

$$\nabla \times \vec{A} = \begin{vmatrix} \hat{x} & \hat{y} & \hat{z} \\ \frac{\partial}{\partial x} & \frac{\partial}{\partial y} & \frac{\partial}{\partial z} \\ A_x & A_y & A_z \end{vmatrix}$$

$$\nabla^2 = \nabla \cdot \nabla = \frac{\partial^2}{\partial x^2} + \frac{\partial^2}{\partial y^2} + \frac{\partial^2}{\partial z^2}$$

Cylindrical Coordinates (r, φ, z)

$h_1 = h_r = 1,$

$h_2 = h_\phi = r,$

$h_3 = h_z = 1,$

$dV = r\,dr\,d\phi\,dz$

$$\nabla f = \hat{r}\frac{\partial f}{\partial r} + \hat{\phi}\frac{\partial f}{r\partial \phi} + \hat{z}\frac{\partial f}{\partial z}$$

$$\nabla \cdot \vec{A} = \frac{1}{r}\frac{\partial}{\partial r}(rA_r) + \frac{1}{r}\frac{\partial A_\phi}{\partial \phi} + \frac{\partial A_z}{\partial z}$$

$$\nabla \times \vec{A} = \frac{1}{r}\begin{vmatrix} \hat{r} & r\hat{\phi} & \hat{z} \\ \frac{\partial}{\partial r} & \frac{\partial}{\partial \phi} & \frac{\partial}{\partial z} \\ A_r & rA_\phi & A_z \end{vmatrix}$$

$$\nabla^2 = \nabla \cdot \nabla = \frac{1}{r}\frac{\partial}{\partial r}\left(r\frac{\partial}{\partial r}\right) + \frac{1}{r^2}\frac{\partial^2}{\partial \phi^2} + \frac{\partial^2}{\partial z^2}$$

Spherical Coordinates (R, θ, φ)

$h_1 = h_r = 1,$

$h_2 = h_\theta = R,$

$h_3 = h_\phi = R\sin\theta,$

$dV = R^2 \sin\theta\, dR\,d\theta\,d\phi$

$$\nabla f = \hat{R}\frac{\partial f}{\partial R} + \hat{\theta}\frac{\partial f}{R\partial \theta} + \hat{\phi}\frac{\partial f}{R\sin\theta\,\partial \phi}$$

$$\nabla \cdot \vec{A} = \frac{1}{R^2}\frac{\partial}{\partial R}(R^2 A_R) + \frac{1}{R\sin\theta}\frac{\partial}{\partial \theta}(\sin\theta A_\theta) + \frac{1}{R\sin\theta}\frac{\partial A_\phi}{\partial \phi}$$

$$\nabla \times \vec{A} = \frac{1}{R^2\sin\theta}\begin{vmatrix} \hat{R} & R\hat{\theta} & R\sin\theta\hat{\phi} \\ \frac{\partial}{\partial R} & \frac{\partial}{\partial \theta} & \frac{\partial}{\partial \phi} \\ A_R & RA_\theta & R\sin\theta A_\phi \end{vmatrix}$$

$$\nabla^2 = \nabla \cdot \nabla = \frac{1}{R^2}\frac{\partial}{\partial R}\left(R^2\frac{\partial}{\partial R}\right) + \frac{1}{R^2\sin\theta}\frac{\partial}{\partial \theta}\left(\sin\theta\frac{\partial}{\partial \theta}\right) + \frac{1}{R^2\sin^2\theta}\frac{\partial^2}{\partial \phi^2}$$

Appendix C: Trigonometric (Trig) substitution

Because of the inverse square nature of the electromagnetic fields, trigonometric substitution is being used more than other integration techniques in this book. Let us review this basic technique here.

In general, trig substitution should be used when there is a radical in form of $(x^2 + a^2)^n$, unless there is an odd power of the variable x in the numerator. Here, n does not have to be an integer, and a is any constant.

For example: $\int \sqrt{x^2 + a^2}\, dx$, $\int \frac{dx}{x^2 + a^2}$, $\int \frac{dx}{(x^2 + a^2)^{5/2}}$, are good candidate for trig-sub.

$\int x\sqrt{x^2 + a^2}\, dx$, $\int \frac{x^3 dx}{x^2 + a^2}$, $\int \frac{x\, dx}{(x^2 + a^2)^{5/2}}$, would be easier with u-substitution.

The first step to set up trig-substitution is to draw a right-angle triangle, and assign the variable to one of the side. Note that the hypotenuse is the longest side of the triangle, so for a function such as $(x^2 - a^2)$, x should be the hypotenuse. Here are some examples:

Functions (examples)	Triangles
$\sqrt{x^2 + a^2}$, $\dfrac{1}{\sqrt{a^2 + x^2}}$, $\dfrac{1}{(x^2 + a^2)^n}$,	(triangle: hypotenuse x, legs a and a) / (triangle: hypotenuse a, leg x)
$\sqrt{x^2 - a^2}$, $\dfrac{1}{\sqrt{a^2 - x^2}}$, $\dfrac{1}{(x^2 - a^2)^n}$,	(triangle with x, a) / (triangle with x, a)
$\sqrt{a^2 - x^2}$, $\dfrac{1}{\sqrt{x^2 - a^2}}$, $\dfrac{1}{(a^2 - x^2)^n}$,	(triangle with a, x) / (triangle with a, x)

Next, assign an angle (of your choice) to be the new integration variable. Then write down two equations. One relates the old variable x to the constant using a trigonometric function. The other equation relates the radical to the constant. Always use the constant.

<u>Example C.1</u>: Evaluate this integral $\int \dfrac{dx}{x^2+a^2}$.

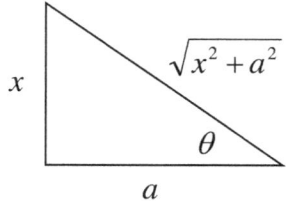

$$\boxed{\tan\theta = \dfrac{x}{a} \\ a\sec^2\theta\, d\theta = dx} \qquad \boxed{\cos\theta = \dfrac{a}{\sqrt{x^2+a^2}} \\ \sqrt{x^2+a^2} = a\sec\theta}$$

$$\int \dfrac{dx}{x^2+a^2} = \int \dfrac{a\sec^2\theta\, d\theta}{(a\sec^2\theta)} = \dfrac{1}{a}\int d\theta = \dfrac{\theta}{a} + C$$

<u>Example C.2</u>: Evaluate this integral $\int \dfrac{dx}{(x^2-a^2)^{3/2}}$.

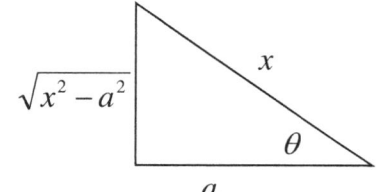

$$\boxed{\cos\theta = \dfrac{a}{x} \\ dx = a\sec\theta\tan\theta\, d\theta} \qquad \boxed{\tan\theta = \dfrac{\sqrt{x^2-a^2}}{a} \\ \sqrt{x^2-a^2} = a\tan\theta}$$

$$\int \dfrac{dx}{(x^2-a^2)^{3/2}} = \int \dfrac{a\sec\theta\tan\theta\, d\theta}{(a\tan\theta)^3} = \dfrac{1}{a^2}\int \dfrac{\sec\theta\, d\theta}{\tan^2\theta} = \dfrac{1}{a^2}\int \dfrac{\cos\theta\, d\theta}{\sin^2\theta} = \dfrac{1}{a^2}\int \dfrac{du}{u^2} = -\dfrac{1}{a^2\sin\theta} + C$$

Appendix D: Wave Equations

In algebra, we learned about plotting a shifted function. For example:

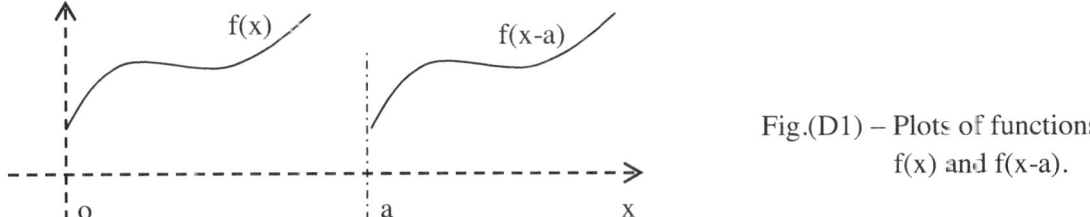

Fig.(D1) – Plots of functions f(x) and f(x-a).

To plot f(x-a) is simply shifting f(x) to the x-direction by "a". If the function is f(x-t) where t is the time, then the function would appear to be traveling to the right over time. To make the units consistent, one should write f(x − vt) where v is the constant velocity of the function in the +x direction. Similarly, f(x + vt) would represent the function traveling to the left at the speed of v. We first show that traveling functions always satisfy the differential equation:

$$\frac{\partial^2 f}{\partial x^2} = \frac{1}{v^2}\frac{\partial^2 f}{\partial t^2}$$ Eqn.(D1)

Proof:
$$f(x \pm vt) \equiv f(u)$$

$$\frac{\partial f}{\partial x} = \frac{df}{du}\frac{\partial u}{\partial x} = f'(u)\cdot 1 = f'(u)$$

$$\frac{\partial^2 f}{\partial x^2} = f''(u)\frac{\partial u}{\partial x} = f''(u)$$

$$\frac{\partial f}{\partial t} = f'(u)\frac{\partial u}{\partial t} = \pm v f'(u)$$

$$\frac{\partial^2 f}{\partial t^2} = \pm v f''(u)\frac{\partial u}{\partial t} = v^2 f''(u)$$

$$\frac{\partial^2 f}{\partial x^2} = \frac{1}{v^2}\frac{\partial^2 f}{\partial t^2}$$

In other words, if the function f satisfies this 2nd order differential equation, the function is a traveling function over time. The direction of traveling is given in the functional form, and the velocity if also given in Eqn.(D1).

Electrical engineers work with frequency and wavelength instead of position and velocity. Parameters ($\omega t \pm kx$) are often used to replace ($x \pm vt$). They are of course the same, knowing that $k = 2\pi/\lambda$ is the wave number, $\omega = 2\pi f$ is the angular frequency, and $v = f\lambda$ is the phase velocity of the wave.

The solution to the differential equation (D1) can be written as $\cos(\omega t \pm kx)$, or $\sin(\omega t \pm kx)$, or exponential functions $e^{j(\omega t \pm kx)}$, each pair of solutions represents 2 traveling waves propagating into +x and −x directions. One could add an arbitrary phase to the solution, but since we have the freedom to choose the origin in location and time, the constant is not needed unless the initial conditions are very specific.

Although the origin function in Fig.(D1) is not pure sinusoidal, but any time function can be written as a sum of sinusoidal waves in Fourier's transform. So the solution sets listed are general solutions of the differential equation (D1).

Example (D1) - A traveling wave is described by P = 50 cos(400t + 20z − 0.25). What is the frequency, wavelength, phase velocity, and direction of propagation of this wave? Show that it satisfies a differential equation similar to Eqn.(D1).

Whatever in front of time t is ω such that ωt is a dimensionless pure number. So ω = 400 = 2πf, or f = 400/2π = 63.7 Hz.

In front of a spatial parameter is always the wave number in that direction. Here, the only coordinate in the function is z. So, k = k_z = 20 = 2π/λ, or wavelength λ = 2π/20 = 0.314 m.

Phase velocity = fλ = ω/k = 400/20 = 20 m/s.

The function is in form of cos(ωt + kz + φ). Propagation direction is in the −z direction.

$$\frac{\partial^2 P}{\partial z^2} = -20^2 P = -400 P$$

$$\frac{\partial^2 P}{\partial t^2} = -400^2 P$$

$$\frac{1}{v^2}\frac{\partial^2 P}{\partial t^2} = \left(\frac{1}{20}\right)^2 \left(-400^2 P\right) = -400 P$$

$$\therefore \frac{\partial^2 P}{\partial z^2} = \frac{1}{v^2}\frac{\partial^2 P}{\partial t^2} \quad \text{which is the wave equation (D1).}$$

In 3-dimensions, the spatial derivative is replaced by the Laplacian, and the differential equation becomes:

$$\nabla^2 f = \frac{1}{v^2}\frac{\partial^2 f}{\partial t^2} \qquad \qquad \text{Eqn.(D2)}$$

In Cartesian coordinates, this can be written explicitly as:

$$\frac{\partial^2 f}{\partial x^2} + \frac{\partial^2 f}{\partial y^2} + \frac{\partial^2 f}{\partial z^2} = \frac{1}{v^2}\frac{\partial^2 f}{\partial t^2} \qquad \qquad \text{Eqn.(D3)}$$

The wave solution in 3-D is now in the form of $(\omega t - \vec{k}\cdot\vec{r})$, where $\vec{k}$ is the wave vector whose magnitude is the 2π/λ, and its direction is the direction of wave propagation. Note that in 3-

dimensional, the function is always $(\omega t - \vec{k} \cdot \vec{r})$, there is NO $(\omega t + \vec{k} \cdot \vec{r})$ term. The "+" term in one-dimension comes from the fact that k and r are in opposite direction, so the dot product becomes negative.

Example (D2) - Given $\vec{k} = -5\hat{y}$. Write a possible wave function to describe this traveling wave.

$$\vec{k} \cdot \vec{r} = (-5\hat{y}) \cdot (x\hat{x} + y\hat{y} + z\hat{z}) = -5y$$

Note: $\vec{k} \cdot \vec{r}$ and $\vec{k}$ are always identical, except the scalar product has the "hat" removed.

Possible solutions (commonly used) are: $\sin(\omega t + 5y)$, $\cos(\omega t + 5y)$, and $e^{j(\omega t + 5y)}$.

Example D3 - A traveling wave is described by $A = 3\sin(1000t + 30x - 40z)$. What is the frequency, wavelength, phase velocity, and direction of propagation of this wave? Show that it satisfies a differential equation similar to Eqn.(D3).

$$\omega = 1000 = 2\pi f$$

frequency: $f = \dfrac{1000}{2\pi} = 159 Hz$

$$\vec{k} \cdot \vec{r} = -30x + 40z$$
$$\vec{k} = -30\hat{x} + 40\hat{z}$$
$$k = \sqrt{30^2 + 40^2} = 50 = \dfrac{2\pi}{\lambda}$$

wavelength: $\lambda = \dfrac{2\pi}{50} = 0.126 m$

phase velocity: $v = f\lambda = \dfrac{\omega}{k} = \dfrac{1000}{50} = 20 m/s$

$$\vec{k} = -30\hat{x} + 40\hat{z}$$

propagation direction: $\hat{k} = \dfrac{-30\hat{x} + 40\hat{z}}{50} = -0.6\hat{x} + 0.8\hat{z}$

$$\frac{\partial^2 A}{\partial x^2} + \frac{\partial^2 A}{\partial y^2} + \frac{\partial^2 A}{\partial z^2} = -30^2 A - 40^2 A = -2500 A$$

$$\frac{\partial^2 A}{\partial t^2} = -1000^2 A$$

$$\frac{1}{v^2}\frac{\partial^2 A}{\partial t^2} = \left(\frac{1}{20}\right)^2 \left(-1000^2 A\right) = -\frac{10^6}{400} A = -2500 A$$

$$\therefore \nabla^2 A = \frac{1}{v^2}\frac{\partial^2 A}{\partial t^2}$$

Appendix E: Standing waves

A standing wave is a result of superposition of two waves with equal wavelength, but traveling in opposite directions. Reflection of a traveling wave superimpose onto itself (the incident wave) is a perfect way to make a standing wave. A few time sequence plots can illustrate this clearly.

Let $E_i(z,t) = \cos(\omega t - kz)$ is the forward wave traveling to the right, and
$E_r(z,t) = \cos(\omega t + kz)$ is the reverse wave traveling to the left, and   Eqn.(E1)
$E_{total} = E_i(z,t) + E_r(z,t)$ is the sum of the 2 traveling waves.

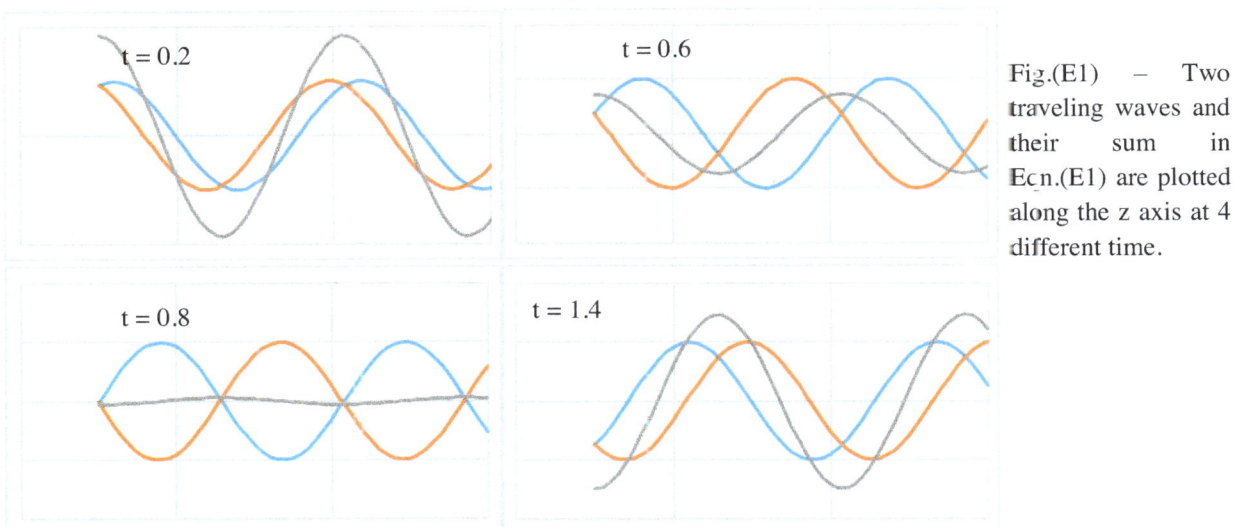

Fig.(E1) – Two traveling waves and their sum in Eqn.(E1) are plotted along the z axis at 4 different time.

Figure E1 shows the snapshots of the waves in Eqn.(E1) at 4 different time. A few observations clearly stand out.
1. With the same wavelength the forward wave and reverse wave have, the resultant sum also has the same wavelength.
2. The resultant amplitude is higher than the incident wave when the reflected wave is in phase with the incident at some particular time, and at other time could be zero if the 2 waves are destructively interfered.
3. The resultant wave exhibits a standing wave pattern, in which there are "nodes" where the amplitude is always zero, and "antinodes" where the amplitude fluctuates the most. This is best demonstrated by overlaying the resultant wave for a few time intervals in the following figure.

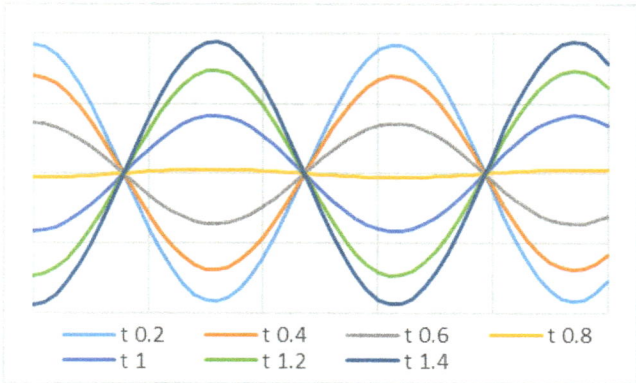

Fig.(E2) – Plotting the resultant wave in Eqn.(E1) over a few time intervals, along the z-direction, clearly showing a standing wave

Mathematically, we can understand this easily by expanding the resultant wave explicitly:

$$E_i(z,t) = E_o \cos(\omega t - kz) = E_o[\cos(\omega t)\cos(kz) + \sin(\omega t)\sin(kz)]$$
$$E_r(z,t) = E_o \cos(\omega t + kz) = E_o[\cos(\omega t)\cos(kz) - \sin(\omega t)\sin(kz)]$$
$$E_{total} = E_i(z,t) + E_r(z,t) = 2E_o \cos(\omega t)\cos(kz) \qquad \text{Eqn.(E2)}$$

Equation (E2) is a generic standing wave equation. It describes 2 independent oscillations: one in time, and the other in space. For a given time, the resultant wave in Fig.(E1) exhibits a cosine curve. And for a given location (in z), the amplitude oscillates over time as illustrated in Fig.(E2). The amplitude of the resultant wave is the sum of the incident and reflected amplitude.

While the differential equation stated in Appendix D is still valid for a standing wave, since it is composed of 2 traveling waves anyway, Equation (E2) has additional classification in its differential form. In fact, one can derive the standing wave equation (E2) from a technique called "Separation of Variables". It is outlined in the following paragraph.

Knowing the oscillation in time and along z-axis are independent of each other, we can write the function $E_{total} = T(t)Z(z)$, where T(t) is only a function of time, and Z(z) is a function of z only. This equation has to satisfy the differential equation (D2):

$$\frac{\partial^2 E_{total}}{\partial z^2} = \frac{1}{v^2}\frac{\partial^2 E_{total}}{\partial t^2}$$
$$T(t)\frac{\partial^2 Z(z)}{\partial z^2} = \frac{Z(z)}{v^2}\frac{\partial^2 T(t)}{\partial t^2}$$
$$\frac{1}{Z(z)}\frac{\partial^2 Z(z)}{\partial z^2} = \frac{1}{v^2} \cdot \frac{1}{T(t)}\frac{\partial^2 T(t)}{\partial t^2} \qquad \text{Eqn.(E3)}$$

The last equation here separates the spatial variable (z) to the left, and the time to the right. In order for them to be equal, the equation must be equal to a constant, independent of z and t.

$$\frac{1}{Z(z)} \frac{\partial^2 Z(z)}{\partial z^2} = -k^2$$

$$\frac{\partial^2 Z(z)}{\partial z^2} = -k^2 Z(z)$$

$$Z(z) = A\cos(kz) + B\sin(kz) \qquad \text{Eqn.(E4)}$$

The constant k is called the wave number = $2\pi/\lambda$. Similarly,

$$\frac{1}{T(t)} \frac{\partial^2 T(t)}{\partial t^2} = -\omega^2$$

$$T(t) = C\cos(\omega t) + D\sin(\omega t) \qquad \text{Eqn.(E5)}$$

The constant ω is of course the angular velocity = $2\pi f$, and it is related to k through Eqn.(E3) so that $\omega = vk$. Putting in the same initial conditions as in Eqn.(E1), the sine terms in Equations (E4) and (E5) would drop out. So the final solution for is once again:

$$E_{total} \propto \cos(kz)\cos(\omega t). \qquad \text{Eqn.(E6)}$$

In most practical cases, the reflected wave has a smaller amplitude than the incident wave. Would the standing wave pattern still holds?

$$E_i(z,t) = E_1 \cos(\omega t - kz) = E_1[\cos(\omega t)\cos(kz) + \sin(\omega t)\sin(kz)]$$

$$E_r(z,t) = E_2 \cos(\omega t - kz) = E_2[\cos(\omega t)\cos(kz) - \sin(\omega t)\sin(kz)]$$

$$E_{total} = E_i(z,t) + E_r(z,t) = (E_1 + E_2)\cos(\omega t)\cos(kz) + (E_1 - E_2)\sin(\omega t)\sin(kz) \qquad \text{Eqn.(E7)}$$

Here, the standing wave pattern is superimposed with 2 standing waves, shifted by 90-degrees in space and 90-degrees in time. The "nodes" where the incident wave and reflected wave destructively interfered no longer cancel completely at all time. The maximum amplitude of E_{total} is the sum of the two waves' amplitude $(E_1 + E_2)$, and the minimum part is the difference of the two $(E_1 - E_2)$. The ratio of these 2 amplitudes is defined as the Standing Wave Ratio (SWR).

$$SWR \equiv \frac{E_{max}}{E_{min}} = \frac{E_{incident} + E_{reflected}}{E_{incident} - E_{reflected}} = \frac{E_1 + E_2}{E_1 - E_2} \qquad \text{Eqn.(E8)}$$

SWR ranges from 1.0 (indicating zero reflection) to infinity (in the case of total reflection). It is a parameter used to indicate how different the materials are, or equivalently, how different the impedance of the waves are across the boundary. Examples of 30% and 60% reflection are plotted in Fig.(E3). The corresponding SWR are 1.86 and 4.0.

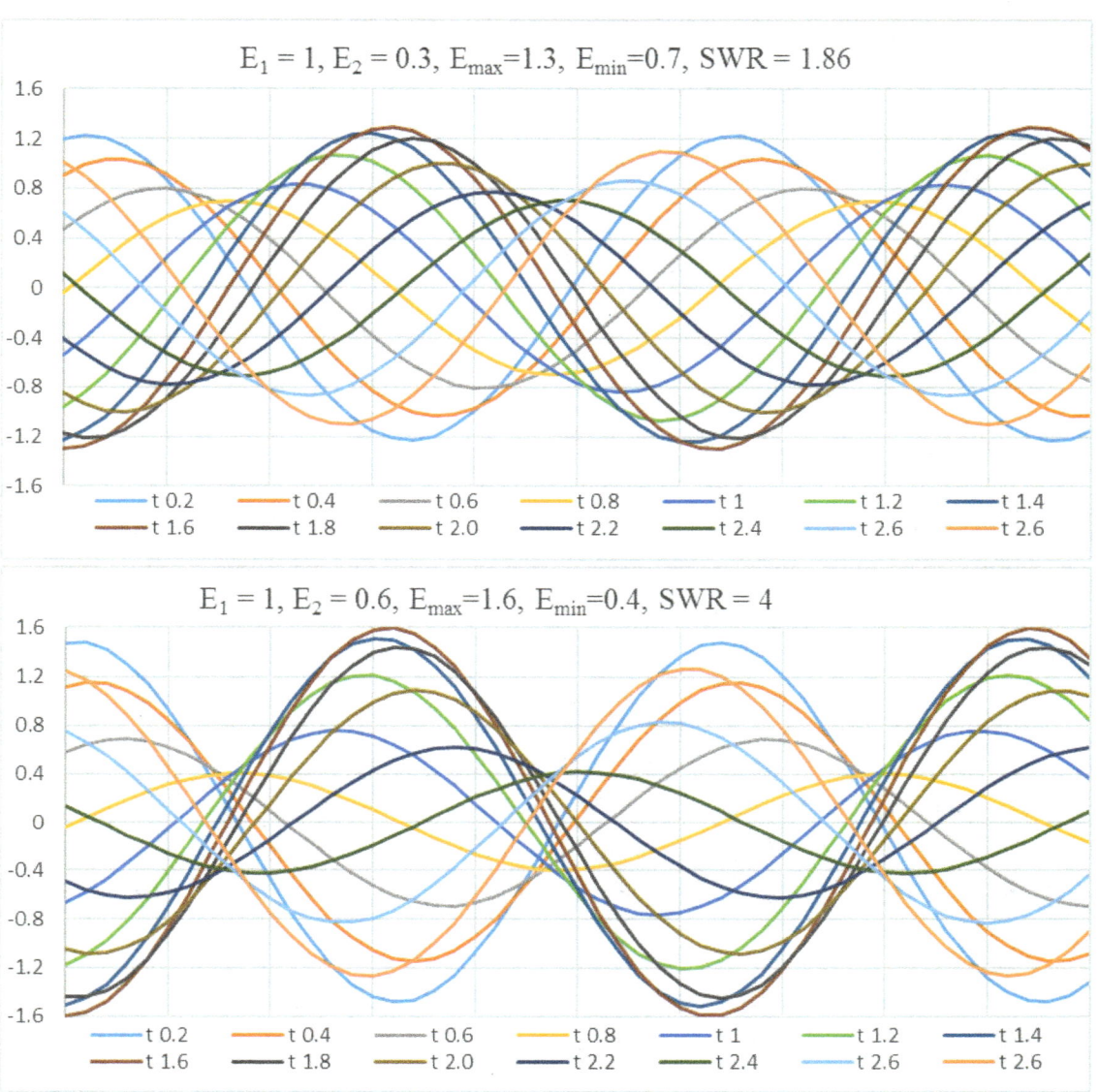

Fig.(E3) – Standing wave pattern for 30% and 60% reflection.

Appendix F: Decibel (dB)

Decibel (or dB) is a logarithmic scale primarily used to describe the power intensity of a wave. Most people first encountered this while learning about sound wave. In that case, dB is a scale to measure the sound intensity relative to the threshold of hearing.

In communication systems, power can range from way over 1 kW to less than 1 µW. Logarithmic scale becomes a standard in everyday RF engineering language. Some of the systems are extremely sensitive and require interference to less than -120 dB. How low is this interference? Let us first define the dB scale.

$$(dB) = 10\log\left(\frac{P}{P_o}\right) = 20\log\left(\frac{V}{V_o}\right) \qquad \text{Eqn.(F1)}$$

where P is the power (of the wave) and P_o is the reference power being compared to. Since power is proportional to square of the voltage (V^2), so the dB scale is 20 times when using the voltage ratio, instead of 10 times as in the power ratio stated in Eqn.(F1). All logarithmic scales in this discussion are based 10.

If the power P is greater than the reference power P_o, log of the ratio is a positive number. In other words, a positive dB number means power gain. Conversely, a negative dB number indicates a power loss.

The term 3 dB is used very often in many occasions. What does it really mean? Knowing log(2) = 0.3, +3dB means the power is double, whereas -3dB means the power is dropped by half. Most of the time people referred to the -3 dB point, or the half-power point.

Some of the common values are listed as follow:

$10\log(2) = 3$,	3 dB = double
$10\log(1/2) = -10\log(2) = -3$,	-3 dB = half
$10\log(10) = 10$,	10 dB = 10 times
$10\log(100) = 10\log(10^2) = 20$,	20 dB = 100 times
$10\log(10^3) = 30$,	30 dB = 1000 times
$10\log(10^{-1}) = -10$,	-10 dB = 1 / 10

<u>Example (F1)</u> – Without using a calculator, estimate what 6 dB is? -9 dB? 7 dB? And -44 dB?

6 dB = 3 dB + 3 dB = (2x)(2x) = 4 times.

-9 dB = -3 dB – 3 dB – 3 dB = (1/2)(1/2)(1/2) = 1/8

7 dB = 10 dB – 3 dB = (10x)(1/2) = 5 times

-44 dB = -50 dB + 3 dB + 3 dB = (10^{-5})(2x)(2x) = 4 x 10^{-5}

Note that dB itself is NOT a unit. It is a scale, a relative number compare to some reference. If this reference is 1 W of power, the unit then becomes dBW, and is an actual unit. If the reference is 1 mW of power, the unit is called the dBm.

$$(dBW) = 10\log\left(\frac{P}{1W}\right) \qquad \text{Eqn.(F2)}$$

$$(dBm) = 10\log\left(\frac{P}{1mW}\right) \qquad \text{Eqn.(F3)}$$

For example, 0 dBm is 1 mW, 30 dBW = 1 kW, -30 dBm = 1 µW. These are real units of power.

<u>Example (F2)</u> – What is 40 dBW? -7 dBm? -26 dBm? And 21 dBm?

    40 dBW = 10^4 W = 10 kW
    -7 dBm = -10 + 3 dBm = (10^{-1})(2x) = 0.2 mW
    -26 dBm = -20 – 3 – 3 dBm = (10^{-2})(1/2)(1/2) = (1/4)(0.01) mW = 0.0025 mW = 2.5 µW
    21 dBm = 30 – 3 – 3 – 3 dBm = (10^3)(1/2)(1/2)(1/2) = (1/8)(1000) mW = 1/8 W